A PANDEMIC IN RESIDENCE

A PANDEMIC N RESIDENCE

Essays from a Detroit Hospital

ELINA MAHMOOD

Belt Publishing

Printed in the United States of America
First edition 2021
1 2 3 4 5 6 7 8 9

ISBN: 978-1-948742-93-1

Belt Publishing
5322 Fleet Avenue
Cleveland, Ohio 44105
www.beltpublishing.com

Cover design by David Wilson
Book design by Meredith Pangrace

In memory of Dr. Zia Ullah

"Let's live in digression. We have no other choices."

—Momtaza Mehri,
"Haematology #1," Frontier Poetry

TABLE OF CONTENTS

MARCH

Testing p.13
Human p.25
O, Rona p.35
Sleep p.41

APRIL

Ventilators and Electricity p.47
Plus, One ☺☺ p.53
The Case of the Missing Feather p.61
The Things We Leave Behind p.69

MAY

Mortician p.77
Holes in the Brain ● p.81
Creatures p.89
Panera p.97
May p.101

JUNE, AND ONWARD

The Dream of a Ridiculous Woman p.109
Post-Colonial p.117
Crawl Me a Slice p.123

Works Cited p.131
Acknowledgements p.133
About the Author p.135

MARCH

TESTING

Sunday

Sometimes it takes the collapse of a system to see things clearly.[1]

I'm three quarters into my first year as a neurology resident. The World Health Organization declared a coronavirus pandemic on March 11, 2020, and Donald Trump announced a national emergency in the White House Rose Garden on the 13th. Just for cultural pin-dropping, the top three songs on this week's Billboard Hot 100: "The Box" by Roddy Ricch, "Life is Good" with Future and Drake, and "Circles" by Post Malone. The Weeknd's new album, *After Hours*, is dropping in a week.

The bars are emptying as we try to make sense of it all. A neurology department-wide Skype meeting is held on Sunday, March 15 (Dua Lipa's "Don't Start Now" has replaced Post Malone's song). Medical students will no longer be rotating in the hospital. Everything that can be done remotely will be offsite. I pace the ground floor at home, microphone muted, listening to the neurology chair's reassuring voice punctuated by sixty chaotic others.

In med school we'd studied the last influenza pandemic, known as the Spanish Flu, which at the end of World War I killed more people than the war itself. We learned the properties of Orthomyxoviridae that allow it to quickly mutate, and that it

1 Through this book, names, diagnoses, and details have been altered to protect patient identities.

was only a matter of time before another flu pandemic came around. COVID-19 (a shorthand for "coronavirus disease 2019") is from a different family, Coronaviridae, of which SARS and MERS are also a part. The argument in the news media about this not being "just the flu" has been baffling—"just the flu" has its own weighty mortality in elderly and immunocompromised populations.

There are two modalities to medicine: immortal and mortal. The immortal part is the education and research that goes into it and naturally draws the human ego and senses: + ssRNA, RNA dependent RNA polymerase, genetic recombination. The mortal part is the craft, which involves patient care and requires the creation of purpose out of purposelessness: how not to have a god complex and still … serve? What's the unclichéd verb for what we're supposed to be doing?

Back to the voices—

"But how are we going to remotely evaluate speech for so many stroke patients?"

"What are we doing to get more PPE?"

The thin, orange-sliced, towering windows that flank the main door to our house open onto a still-wintered suburban Detroit, the darkening sky darkening ugly patches of snow with it. Is it real or is it not? How much of this is spectacle and how much of it is real and how real? As much as a mind or more, maybe greater.

Monday

My teaching hospital has started taking drastic measures

before many others in the nation, and is doing a remarkable job. However, it is doing so outside of what should have been an earlier, federal government-initiated preparedness. Rather than this being a streamlined nationwide action, it's been entirely too fragmented. Without having had a simulated health response, we've been left to on-the-moment trial-and-error planning. The information pouring in is saccadic (vertically so).

The screening process began misshapen today. Lines wound all the way out to the parking structure under a dark morning sky, and mask-less hospital personnel waited an hour just to get into the building. Apparently, this is being done in other hospitals in the county. The unfortunate part is that non-hospital personnel have been joining these lines thinking they're in line for COVID-19 testing only to find just an oral thermometer and a few screening questions after waiting for hours.

Tuesday

Shaking hair and shedding layers of melting snow, we leave our coats to mold on the coat rack in our stroke unit workroom. Everyone is in scrubs now—it makes sense to wear scrubs. I log into our electronic medical system to chart review my patients in our rapidly dwindling patient list.

All admitted patients suspected to have been infected were initially having their labs sent out. The country did not have adequate equipment for testing. The pandemic is revealing the problems that arise from virtually zero home production of essential supplies. We have testing available now, the first hospital in Michigan to start in-house testing, but measures are still progressing slowly. Trump's press conference today was a joke (as if anything else was to be expected, with the CDC budget cuts and the disbanding of the pandemic response team in 2018).

We're no longer admitting patients who shouldn't have been admitted in the first place. Admissions that end up doing more harm than good in the long term—a defensive form of medicine that I've often questioned. Instead, we're triaging people who don't need to be in the hospital, where they're at a higher risk of developing an actual infection or problem.

We're dealing with acute illnesses and finding the necessary time to attend to their conditions without feeling like humans that would have been better served as robots. People are presenting to the ED for issues that should be addressed in an outpatient setting but can't afford to due to lack of insurance or social support. HIPPA violations are lifting enough for us to start an effective means of tele-medicine, something discussed for years but only now coming to fruition. I don't want to just be another meme (in what is quickly becoming a very robust meme game), but this feels like what being a physician should be like: the chance to feel meaningful.

Wednesday

Viktor Frankl is best known for having studied and written about the importance of meaning. An Austrian neuropsychiatrist and Holocaust survivor, he developed logotherapy: the therapy developed from the concept that finding meaning in life is a human's primary purpose. His memoir, *Man's Search for Meaning*, recounts his time in the concentration camp, and details how even in the most dismal human conditions he and the other prisoners were able to find spiritual integrity—that in fact calling forth spiritual integrity in acute duress provides the opportunity to achieve greatness.

Frankl flips the idea that meaning is found in human agency in creating a future to say that whatever future unravels has the potential to have meaning created out of it. Perhaps not surprisingly, for Frankl meaning is inexorably linked to suffering:

> "Long ago we had passed the stage of asking what was the meaning of life, a naïve query which understands life as the attaining of some aim through the active creation of something of value. For us, the meaning of life embraced wider cycles of life and death, of suffering and of dying."

I'm trying to avoid contact with my family and isolating myself in my room when I'm home. Despite the movie *Contagion*'s uncomfortably parallel narrative, the announcement of the pandemic has been surprising for the U.S., where order and routine are akin to godliness. The standard, strictly ordered medical system has been replaced with confusion. Despite daily email updates, we are still uncertain of what is being done and needs to be done. Rules about what counts as safe protection and possible treatment modalities are changing on the hour.

Thursday

We had a patient, Mr. Tracy, transferred to us from vacation in the Caribbean in the middle of the week. He'd been on a cruise, developed a hemorrhagic stroke, a brain bleed, and was flown back to us for treatment. I wasn't able to see him for over an hour when he initially arrived via air ambulance, because I couldn't figure out where to get protective gear.

While pictures of people in masks and gloves in supermarkets have been circulating on social and news media, we still have a crippling lack of masks. There are usually masks available outside patient rooms, but visitors had been taking them in the days leading up to the actual announcement of a national emergency and we've been left with an even shorter supply than anticipated. However, one of my seniors saw this wonderful display of human comradeship and had the presence of mind to stock a supply for us in the workroom ceiling. It was here that I was finally able to find a mask before going into Mr. Tracy's room. We're told to keep PPE, including surgical masks and N95's, in labeled brown paper bags to reuse for a week.

It's cumbersome maneuvering in protective gear. I spend an inordinate amount of time trying to maintain OR-like sterility without any of the basic facilities to do so. The N95 has a musty smell that takes a second to adjust to. Once inside, I was only able to make out some of Mr. Tracy's frustrated history through my yellow plastic gear. His Broca's area, the part of the brain responsible for speech, had been affected by the stroke. The result of this was that he could understand what I was saying, and knew what he wanted to say, but the words just wouldn't come out right. Broca's aphasia is an understandably frustrating condition, but Mr. Tracy's situation has been made even worse by the timing of his stroke. Visitors have been banned (not to save masks, there really aren't any to save, but to facilitate social distance) and that means Mr. Tracy has been physically isolated from his spouse and family. He can only reach out to them through fragmented air waves. I got his wife on the speaker to help calm him down: "It's okay honey, I know, I know."

Our inpatient team decided to split itself, with half of us rounding from home while the other half comes in to work in person. Mr. Tracy's brain imaging was concerning for an aneurysmal source of the bleed, so I put in the neurosurgery consult from my laptop in bed, and called the resident on call. Cross-legged on my white comforter, I then called his wife.

Friday

There has been a steady uptick in research, most of it originating from the Wuhan province in China, where the pandemic began. A small, non-randomized clinical trial by Guatret et al. (2020) studying hydroxychloroquine (an antimalarial) and hydroxychloroquine with azithromycin as treatment modalities has been published. I've heard through word of mouth that some physicians in New York have started using hydroxychloroquine prophylactically. There are also studies and theories floating around that fresh air and sunlight helped mitigate the influenza and SARS outbreaks in the past and that a change in season may ease this pandemic's acceleration, though there's little legitimate research to back this claim either. We're being encouraged to wear surgical masks at all times, something that had been debated in the weeks leading up to this.

Dr. Peter Safar (1923–2003), was an anesthesiologist who is credited with having developed CPR in the 1950s. CPR is a part of "code blue" activations in hospitals. Safar is also accredited with the ABCs of resuscitation, the establishment of intensive care units (ICUs), and with setting up the first emergency medical system. Anesthesiologists used to run ICUs; they've been replaced in that capacity by pulmonology critical care doctors.

Internal medicine residents are being pulled from various rotations into the emergency department to replace their exposed associates. Our ICUs are usually at full capacity on a normal day, so our administration is preparing for the worst and creating more ICUs. Entire floors are becoming dedicated to COVID patients. It is Friday now, and overnight there were approximately ten codes spread across the various COVID floors. "This is a code blue alert, P as in papa, 566." The hospital's first COVID-19 death has occurred.

Saturday

A few hospitals in our county decided not to heed the WHO notices. Even though countries like South Korea (which learned its lesson after MERS) are doing a great job with extenuating the virus by testing people, the CEO of one of our local hospitals, let's call it St. Bartholomew's, claimed that "testing won't help in treating patients." St. Bartholomew's admin also didn't heed the Surgeon General's advice to cancel all elective surgeries. So now, St. Bartholomew's decisions have led it to the brink, ready to overflow with infected patients. This pandemic has shown what complete lack of common sense can look like—the initial rush for toilet paper in our country becoming a world mockery.

There's a meme circulating (as a side note, it's bizarre how that word, "meme," transformed from Dawkins's Selfish Gene into its current formulation) that I've found striking. Say we come up with a vaccine. Fifty or seventy or a hundred years down the line, this pandemic will no longer be fresh in living memory. We'll relive the same story of people refusing vaccination. That's the struggle with prevention altogether. It's hard to imagine something you've never seen—it requires a modicum of belief in history and science.

But if we go the way of logic, it can be taken back a step further, even before vaccination. A lot of these viruses originate in animals, and it is possible to break the transmission from animals to humans. We can start prevention by monitoring diseases in animals and having an established dialogue between the health system and veterinarians serving high-risk areas where the disease is most likely to make that interspecies transmission. However, this requires a lot of effort with often little to show for it: the absence of a pandemic is a hell of a lot less discernible than a pandemic. Unfortunately, especially in this case, memory necessitates repetition.

Sunday

Medicine teaches you how to approach emergencies with urgency but without panic. I started my residency on the internal medicine floors, where mornings would start with my coffee masked by the smell of human excrement from open doors as nurses made their morning cleanups of their patients. I'd developed a soft spot for one of the patients who had come into the hospital two weeks previously with complications from recurrent prostate cancer. I'd gone to evaluate him when he'd been initially transferred to us from the ICU. He was angry at being moved to another floor for what felt like the tenth time. I knelt down next to him and his anger transformed into despair as he started sobbing. He was an engineer and had been trying to approach his sickness with what he called a "mechanical outlook," but was struggling.

Cancer puts you at a risk for blood clots, so you will often see cancer patients on anticoagulants, or "blood thinners." I re-started him on Lovenox, one of the blood thinners, before I realized that his GFR, a measure of kidney function, was low and that he had required a lower dose of medication. I panicked and called the on-call senior, who came into the workroom from one of the other four floors he was responsible for. He helped me calm down and we'd called the pharmacy to make sure they were aware. The pharmacist was eating something and casually responded that she'd already adjusted the dose, and had been about to call me about it anyway. The modern-day tertiary care medical system has a lot of safeguards involved, and the holes must align in our ever-cited Swiss cheese model for a legitimate error to occur.

No wonder already-struggling rural health care centers have been closing down at such a rapid speed. They cannot maintain the medical structure that has become the norm. The few rural centers that have survived are understaffed with doctors that work close to seventy-two-hour shifts, without a department of pharmacists or an army of experienced nurses. The nurses in our tertiary center perform venous blood draws and insert cannulas—I don't remember the last time I had to place one. I can't imagine normal hours without those safeguards in place, never mind having to work for three days while doing blood draws and double and triple checking medications.

Four months into my intern year I was on the pulmonology floor with the same senior. Four months had made me comfortable with the medical structure, the environment, and myself. I knew what needed to be done before calling my senior: patient with chest pain? Go and assess the patient; where is it, when did it start, is it tender, does exertion make it worse, are you having palpitations or nausea, are you feeling short of breath or diaphoretic, does the pain radiate to your left arm? Check vitals for hemodynamic stability, get a stat EKG and blood troponins. My senior saw me come back to the room one day after assessing a patient with dropping oxygen levels and drawing an arterial blood gas: "Look how much you've grown, remember when you'd been freaking out about the Lovenox? You could run the floor on your own now." Definitely an overstatement, but I remember that as the first external marker of finally getting the hang of things, of finally being able to face the emergent with a sense of calm.

You learn to distill stillness into fire, and the current circumstances are an external reflection of that: there's a dead calm in the hospitals. The corridors are more deserted than I've ever seen them, visitors banned, working staff rushing past each other, trying to maintain distance, yet at the same time our beds have never been so full.

Stories from those on "the frontline," mainly those in the ICU and ED, where the incidence and mortality of those who have been infected is climbing, have been circulating through the hospital. It's the young deaths that take you by surprise. Most of the hospital residents are in their late twenties and early thirties, and although we initially considered ourselves protected, that's not necessarily the case. I don't think any of us are actually contemplating that we may die from this, that worry is still reserved for others, but it gives you pause.

Dozens of doctors in Bulgaria resigned a few days ago due to lack of protective gear. It is Sunday now and I left my twenty-six-hour weekend hospital shift to drive home on a completely deserted highway with The Weeknd's *After Hours* in my ears and Viktor Frankl on my mind: having been is also a kind of being, and perhaps the surest kind.

HUMAN

Nominative determinism

The earliest memories of my father are blue: waiting with my mother outside the medical residents' apartment building in Detroit, where I was born and lived for the first five years of my existence before we moved out to the suburbs, as he approached from the hospital in seafoam scrubs with his characteristic brisk walk. It's a memory I believe to be mainly factual, because it's subtle enough not to have been recounted into fabrication (subtlety is a good measure for accuracy; the more subtle a memory is, the less likely to be recounted and thereby less likely to be blemished). The created part of the memory: an infant me, unable to say "Baba," reaching out to "Babi," is cute enough for me to edit in.

Fast-forward twenty-some years later and I'm back in Detroit, where my parents completed their residencies in neurosurgery and psychiatry, in similar scrubs, with a brisk tongue (comma, pen). I'm little more than halfway through my intern year in neurology, blue silenced under layers of snow, in a universal system deeply flawed in many terrifying ways. A snippet from Samuel Shem's *The House of God*, written during the Watergate scandal and Nixon's resignation and a loosely fictionalized biography of five residents' progressive disillusionment with medicine as they make it through their intern year at Harvard's Beth Israel Hospital, reflects some of my mawkishness:

"...that's modern medicine."
"You're crazy."
"You have to be crazy to do this."
"But if this is all there is, I can't take it. No way."
"Of course you can do it Roy. Trash your illusions, and the world will beat a path to your door."

It's a battle fostering a non-delusional yet ideologically sound spirit in this structure. I regularly find myself either becoming a part of or folding into myself against the onslaught of human hubris (an image borrowed from a cousin doing his PhD in the humanities, apparently a similar realm). What of a life, save for what, to what end, to what purpose? I'm suspicious some days. There's a staggering lack of discussion, outside of palliative medicine at least, about what a good life or death means, about whether human progress can be pinned down to a matter of quantitative years.

In *The House of God* there's a mention of an international medical graduate's indecipherable transfer notes. And there's another paragraph in the book that caught me by surprise: "A Hindu anesthesiologist pumped oxygen at the head of the bed, looking over the mess with a compassionate disdain, perhaps thinking back to the dead beggars littering dawn in Bombay." Despite, or perhaps because of that, I loved *The House of God*—written under a pseudonym by psychiatrist Stephen Bergman, and based on his own internship experience—for its authenticity. It stands apart from other medical stories and their elaborate savior complexes. This was the first novel I'd read where the political drama of the system was explicitly hung loose. How did the protagonists of this book look at their out-of-country colleagues? What creation stories did they attribute to them? Memories wiped out for easily digestible clickbait.

How many branches will it take before I, too, see them as other? My grandchildren, their children? How many memories will it take before "we" are finally erased—until Lahore is seen as a singular vision of third world-ness? The Lahore I knew growing up was proud. Erect and bejeweled. It was cosmopolitan and otherworldly—American and Punjabi and British and French accents moving to the lull of a honey-tongued Urdu. A bubbled part of society, certainly, but it's all I've ever wanted of it.

Subtlety is more damning than forthrightness. Microaggressions are thus peculiar creatures: how do you handle something whose existence is questionable to begin with? More often than not the answer either ends up being a catastrophizing, exaggerating, and philosophizing of monstrous proportions—or, alternatively, silence. The two often feed into one another. Collected over decades, enough silence putrefies and snowballs at an alarming rate: I should have defended myself, my family, my nation, my religion, us. What is "us" even? Variations of us are often just as belittling and condescending: third generations derisive of second generations derisive of first generations. I don't know at what branch point "us" becomes "them." What I am aware of is the fear of us becoming them—once you've been us-ed, the desire to hold back kicks in: a centripetal force sucking the branches back into the trunk, creating its own kind of us. But there was never an us, just a gut. One gut, and the us comes slamming down, bubbling us together into a frightened corner. I am not us, I am us, weak, you should have said something. The parting words—you should have said something.

Requiem of the subtle

According to Yeats there is poetry of the memory and poetry of the imagination. The latter may be more sublime than the former, but the former is a necessary precursor to the latter. I've often wondered what memories my father lives for—his family, his childhood, for the scent of a Lahore that ceased to exist the moment he moved away. He is a stoically soft proverbial girl-dad to three daughters. It was a rare moment at dinner in Lahore with my baby sister that he recounted how he first moved away after graduating from medical school: I looked out the airplane window and just knew; I knew I was leaving forever.

Leave, forever, go back, an account at its best, here's the rosary, I'll hold it while you climb back, here's the story, I'll hold it back, here's the end, I'll hold it back, here are the borders, we'll hold them back. The freedom to roam, the freedom to be hindered, visions of a monolithic Muslim—the density of ignorance, globe-trotting-border-demolishing delusions, we are who we are, here, bolded, we'd crossed borders long before there were borders and long after, before and after some center became an east and some center shifted west, across some dense beginnings and dense ends, we are who we are, bolded.

And what of them, that is, why of them, these bolded borders? Borders are necessary (are they necessary?), borders of the self, family, nation, etc. Now that they exist, they exist—though the question of their porosity remains: do you gash them open, or slit them shut? Or perhaps the pores are like gaping fish mouths: their O-ness dependent on various extrinsic and intrinsic factors. Me and the demolishment of me in a state of perpetual fluidity. Plop.

First-generation immigrants view immigration differently from their second-generation progeny. There's an unquestioned linearity of progress that comes for the first, and for that reason immigration can be a little less existentially troublesome for them. People move to the West (and I say this with Edward

Saïd in mind) for an expected and well-perpetuated fiction of progress. Second-generation children, on the other hand, are projected to live in a dream that they never dreamt. They are viscerally unaware of the time-space from which their thought-bubble-being-of-an-existence originates. That original idea of linear progression, of power, can come back to haunt them in an occasionally odd, sometimes paradoxical, struggle.

In my imagination I'm somewhere along or below or beyond second generation. I've moved back and forth enough to have called both countries home, but somehow still found myself struggling along the third-culture kid spectrum. I'd created an ambiguous, emotionally charged struggle with belonging: an elaborate, delusional ideology—of us and them, nationhood, belonging, home(bound), Salman Rushdie's *Fury* not furious enough—that had me run back to Lahore for medical school a few months after graduating from Ann Arbor.

While most of the world goes straight into medicine after graduating from high school, in the U.S. you still have to go through undergrad before medicinc. This is somewhat counterintuitive, because students enter medical school at the back end of malleable brains. This is largely due to the *Flexner Report,* a book-length survey of the state of medical education released in 1910, that cited Johns Hopkins—fancy almost since its inception, and having implemented additional prerequisites—as the "model medical education." In its wake, the Rockefeller Foundation provided fifty million dollars to all the schools that adopted the Hopkins models. The *Flexner Report* increased the qualifications needed to get into medical school and ended up closing a lot of schools, including almost all of the historically Black ones. This report is sometimes marked as the beginning of the divergence between what we think of as medicine and public health.

Anyway, despite my parents' advice, I fled that fall after graduating. When in Rome, be Roman (Greek? No). No, it's a choice, our own fault. And so, I went back—swapping one set of assumed unbelonging for a different set. Places are like people: packages of good and bad. You get to choose what set of good and vice fits your wants. As for the bad: if it's not your religion that's an issue it's your race, if it's not your race it's your gender, if it's not your gender it's your social class, ad infinitum. Different pulls sucking you in—trunk bared. I never did end up finding "us."

Creation story

One comment, one tilt, hurtles into our creation story. It's raining as it can only rain in Lahore in the summer and the fan's slicing the lemons and bathing us in sweet acid rain. The rain-soaked earth attar pouring through the iron-latticed grill protecting the open windows. I'm in my maternal grandmother's guest room, what was once my mother's room, with posters of *Grease* and ABBA long since peeled off, laying back on the Styrofoam mattress slice slice slice. We were only here, only everywhere.

Per one of my medical school professors, for our people, for people of the heat, all maladies begin in the gut. I was in Lahore when Trump was elected. There was a silver lining to this. Whatever he was (I'm unsure why that falls into the past tense, but it is—past), Trump wasn't prejudicially dishonest (or his dishonesty was so thinly veiled so as to basically be equivalent to an unintended honesty). And for all that's worth, it suddenly illuminated a collective narrative of microaggressions. Until his election, I hadn't realized that my friends had gone through similar experiences growing up: PTA meetings urging bilingual children into special ed, confusing bilingualism with low IQ. I started reading classics before the age of nine to disprove my stupidity; *David Copperfield* my first attempt at a self.

Here's the self; I'll trumpet it. At least when an opinion is plainly voiced ("go back to where you came from") there's a way to contest it without being denounced as paranoid. You can argue with the illogicality of it or make whatever rejoinder that people who uphold that claim make. It's a different, and in some regards more honest, struggle than the never-ending struggle that comes with the ambiguity of microaggressions, of fake accents, of littering beggars.

I'm wary of overt populism and the way things end up going both ways; the inconsistency of it. Everybody is guilty of something, of some kind of aggression, and that makes defense of anybody or anything questionable on some level. It's the problem with being human. That being said, silencing yourself in the name of consistency is a rather dangerous path.

One comment, one tilt, one accent, hurtles into the creation story of "us." The past is so distant, we've reverted to the original narrative, as if we'd never left Detroit—the gap in history a panting lisp. The pot's burning the lemons over the electric stove in Detroit—and I spoke too soon—quiescent-soul-vomit bubbling to anything brimming of anything. I never did end up finding us. Or maybe the "us" I ended up choosing was American. Second/sur generation, Pakistani, left. I flew out of Lahore for New York after I graduated from med school, and this time, there was no forever in the past, no back in sight.

Grounding

The first place my father landed in the U.S. was Scarsdale, New York, where my grandfather's brother, known as Chacha

Jaan, practices as a pediatrician and where my paternal great-grandmother is buried. My father would live there for a few years before he and my mother started their residencies in Detroit. Fast-forward twenty-some years later and I'm back.

I was on an internal medicine inpatient service and went into my patient's room to pre-round. A large family greeted me—a woman with a puffy pink jacket in the corner saw my name tag and excitedly asked, "Are you related to Dr. Mahmood!?" He'd operated on her five years ago. "Your dad saved my life! I like him, he's serious and about his business."

Between the paperwork and insurance dealings and hubris tide and redundant EMS system, there is still a life. A matter. I still have a hard time digesting those words: saving lives. The words are heavy and loaded and I'm suspicious of them. Too strong, too power-hungry, too fluent, too easy, too unthought. This is a doing—saving, another matter. If it's a matter of trying though, my father's done more than his share by quite literally dedicating his life to his work and family. If it's a matter of trying, he's tried to create honesty in a hyphenated existence for us.

I've devoted much of my mind to lamenting immigration from both ends: why do we leave, why do we return? Power struggles, egos, a desire, we try. Power, a continuing centripetal force throughout history, doesn't have to be inherently evil. Equality in an unequal state, tipped as we are into empire. A down payment for our own deaths, but how else could it be? This is how it's always been. And yet, there has to be some festering hope that power can be held with grace. We no longer belonged there, but humans never really belonged anywhere, belonging a ruse. So, we try.

We did not leave now, or recently, or in our personal journeys. In a twist of an Ocean Vuong poem: we left so long ago we can no longer see when or where or why, all we know is that it was done, has been done again and again. Virginia Woolf in *Orlando* described the dissolution of illusions so that others may be built in their stead. I have changed and swapped so many illusions, illusions of home, illusions of whole ideologies, that I wonder how I still am at the other end of things. Memories as backdrop, a reservoir to sink into as beliefs swap in and out. Blue as we watched, blurred in and out.

Whenever I see my father in the fluorescent undertones of the hospital he calls a second home, brisk in place, eyes bright, I am reminded—I can be better, human, a human. It amazes me as much as it breaks my heart to see how happy it makes him to see me here (I'm the prototype that finds pathos in poignancy). A memory's only as good as we make it. We were only here, only everywhere. He's often said a thousand lives over he'd still choose this. A home foregone, a home embedded. Again and again, an eternal recurrence—I'm still striving for a similar amor fati.

O, RONA

Scratch that. All on the floor.

It's the evening of March 24 and my neurology co-interns and I got notified that we're being "redeployed," that is, pulled from our current rotations to make a new COVID floor. Not gonna lie, I was scared. Not gonna lie, I wavered. I'm going and I can't believe this sentimentality is spewing from my mouth—but damn, it really feels like we're going to war.

I moved out the day following my draft (in keeping with that ongoing metaphor). My parents are in a concerning age bracket, with associated comorbidities. I emailed the program coordinator for some sort of temporary lodging. Until then, I'm living with one of the other interns. I packed some of my belongings and random house supplies in case I end up needing to rent a place. My mother filled a laundry basket with fruit and entirely too much food. I carried it out of the garage into the dark airless night and put it in the backseat of my silver baby SUV. No one knows how long this is going to last.

While we're constantly being told—by sources on social media, by self-help gurus—to focus on the present, I appreciate Frankl's emphasis on what is an imagination of time: a circular creation of looking into the past from a future moment to create the present. He writes about how meaning-making is intertwined with the future. If we are able to envision a future from which we can retrospectively make sense of the past, a meaningful present is created. Moreover, make your life about meaning and side-wishes might ensue.

G'morning and onward

The day started off slow with a Skype business lecture on SARS-CoV-2. MERS, NL63, SARS were postulated to have had their animal source in bats. OC43 had its source in cows. It is questionable how SARS-CoV-2 made its florid entrance with the first documented case on December 8, 2019 in Wuhan. WHO had declared a global alert by January 30, 2020. The case fatality for the seasonal flu in all ages is reported as 0.1% and COVID-19's is relatively higher at 2.3%, disproportionately affecting those above the age of sixty and with the following risk factors in decreasing order of risk: cardiovascular disease, diabetes, and chronic respiratory issues. The R0 value, or basic reproduction number that indicates how many people a sick person will infect on average, is postulated to be 2.5. The higher this number is the more contagious the infectious. For comparison, these values are 16 and 6 for measles and smallpox, respectively.

It took a while to set things up after the lecture, but once the protective gear (what they were able to get at least) was in place, the floor opened, and admissions started rolling in. We were taking patient histories on the phone before walking in hazmat gear into the patients' rooms with the attending staff physicians.

The U.S. officially has the most COVID cases in the world now. (Must we really keep capitalizing that? I can't anymore.) We got an email today from our system informing us that we'd had sixty covid patient-related deaths in Michigan and two of them were our fellow health care workers. No names were offered. Every time I start feeling like I'm being overly dramatic something happens to trigger a reality check. Physicians and health workers have been talking to their partners about code statuses—meaning, whether they would want to have CPR and intubation.

Surgery residents have been redeployed into the ED for swabbing patients. The orthopedic surgery residents are now running their own covid floor (much to the amusement of almost everyone in the hospital). Urologists, neurologists, radiologists, dermatologists, and ophthalmologists are all being redeployed. The hospital has basically become a covid center. There's worry that Detroit and Chicago are the next New York, which has gotten the worst of it in the U.S. I'm still trying to figure out what my living situation will look like for the next month: whether that means living in a hotel or being able to find my own apartment.

There have been unique social challenges with this pandemic. Our hospital is located in downtown Detroit, and the less socioeconomically privileged are the ones who suffer the most. I had a patient with cognitive delay present after her family member had kicked her out of her house because she "had the virus." She'd walked to our ED and tested positive. I'd waddled into her room to see an affable middle-aged female with clean, round black glasses. Once she was weaned off of oxygen and medically stable, we didn't know where to discharge her. Most of the rehabs, nursing homes, and shelters have been struggling with community-spread coronavirus and understandably are no longer accepting people. Even if they were, the chances of getting others sick is too high. Most of the deaths have been in nursing homes, wiping out entire groups of people. Where do you send them? We eventually discovered, in typically confusing fashion, that the City of Detroit and Detroit Continuum of Care had developed a process with hospitals to help arrange for housing for patients during this pandemic. We were finally able to arrange safe discharges.

Science et al.

There are certain inflammatory markers that have been associated with a greater risk of mortality: increased D-dimer, a protein fragment of dissolved blood clots, and lymphopenia, a decrease in a specific kind of white blood cells, were the most important ones found in a Wuhan study by Fei Zhoul et al. They found an increased D-dimer in those who had died. We've personally found trending ferritin, a type of blood protein that carries iron, most significant in these patients on the floor.

We've been using an algorithm to categorize patients as either mild, moderate, or severe. The moderate and severe ones are getting Plaquenil (hydroxychloroquine) or getting enrolled for Remdisivir (I heard about some absurd conspiracy theory with Bill Gates starting the pandemic just for Remdisivir and you've got to hand it to humans for a twisted sense of creativity in times of duress). Sicker patients in the ICU are getting Tocilizumab, an IL-6 cytokine inhibitor. We're also treating patients with elevated procalcitonin, a peptide precursor, with antibiotics to prevent them from superimposed bacterial infections.

There were initial studies that steroids didn't help. We know steroids aren't a go-to in influenza due to concern about viral shedding, but the trend is changing. These patients are hypothesized to have something called "cytokine storms." Cytokines are small signaling protein molecules that take part in various bodily functions including inflammation, cell differentiation and growth, and immune regulation. A cytokine storm is when there's a superfluous output of cytokines causing an overwhelming immune response. Patients will be on the floor on their nasal cannula oxygen and suddenly start going into acute respiratory distress syndrome, also known as ARDS.

The thought is that the cytokine storm causes the ARDS crash. The initial threshold for starting steroids was if a hypoxic patient was requiring four liters of oxygen through their nasal cannula or if they met the standard early ARDS criteria; however, I left at the end of the day with the discussion of whether that threshold should be moved to two liters of oxygen to prevent the cytokine storm before it occurs. A week and a half later it's changed to anyone who's hypoxic and needing any supplemental oxygen should be started on steroids. This is slightly counterintuitive, treating acute lung infections other than asthma and chronic obstructive pulmonary disease with steroids. However, it seems to be working and the data from our hospital has been promising.

A week after having been on the floor, we finally got a schedule for what the coming month may look like, and the co-intern I was living with was moved to another service. I moved out into a hotel owned by a family friend. I haven't had a day off in over two weeks and days are starting to melt into one another, sleep another matter.

SLEEP

"The nights are the hardest. But then the day comes! And that's every bit as hard as the night. And then the night comes again…" –Rachel Green, Friends

You dreamt that I had a heart and I dreamt that I lost it so we reconciled both I landed on the conveyor belt escalated from airport terminal to baggage claim as baggage the furious prayers of now now now still resonating differently less purely less angrily now now now you didn't come home so I didn't either. I left it in the trunk or boot or whatever you'd like to call it, the goats being herded for zabiha or sacrifice on Bari Eid the recreation of Abraham and Isaac's story heads marked for the red with haloed pink bands. There's a picture of me with one that looks like an alpaca in one of the many baby albums in the basement. I march in the way of being away from where you think this is going, away from practice to fanciful theory. You can see science fall apart in the way of medicine. I'm still in the way of practical. Still fanciful. We'll march to this tune together—

In the patient's room, banging open curtains, letting sun wash away delirium. Standing still watching the attending examine the patient, a cup dithering on my piriform cusp, waiting it out, it so rarely visits anymore, this deepness of purpose. If I had to describe a living will, it wills itself through a trap in the chest, substernal, up through the sternal notch, until it raises up and lodges like a fishbone in the piriform fossa and you wish you could aspirate it, tracheostomized, and make a moment, any moment, worthwhile. I remember everything in that moment, everything that I'd deleted to make this the way I want. And then the loss of it all. The happiest are those who forget. A false happiness in some phrases. Happiness was when we could hold our pasts, when they were once small enough to fit in our palms. I walked out of the hospital and you'd become someone I used to talk to in undergrad. The rest, the leaves dragging time in their wake, spilled over as compost. In my imagination it's already past Thursday and it's too late now.

Time only becomes a reality with ends, and only something you love can end. Something really must have ended in the heart of humanity during the Industrial Revolution then, when time became a big deal and was set to Greenwich time. She was shifted from the hourglass to faceless mechanical clocks, and then painted with the face we associate with her today. She was pendulum clocks and pocket watches. Then from frequencies of the visible earth she delved into the invisible frequencies of the atom, cesium, that sky blue element to which now she is linked. Atomic time. Think of her for a second and think of the brilliant blue sky spinning nine billion times. That's a second. That's time.

Humble the Poet writes that as kids we don't do things for goals and as we get older everything becomes about goals. Even free time becomes a goal. That adds an unforgiving rigidity to life. When we were younger, hours had been fluid. Sleep wasn't for sleep and getting up wasn't for getting up and school wasn't for school. Things were organic, grounded. We went to school came back talked to the people we had just parted ways with until we started doing homework until we fell asleep until we woke up. The existential exhaustion that comes with purpose hadn't set in. Now, driving home from the hospital, it takes every ounce of energy to call my soul splits.

I don't think it, whatever "it" is, ends at the deep end. It ends where sleep became a goal—its lack of and need of and crave of and mystery of. Obsessions with the dreams it creates and how the obliteration of those dreams sets the trap for the death of something so pure it can only be related in metaphor: a trapeze artist facing his purposelessness. It ends where we let the continuity in the narrative break. It no longer fits in our palms, it spills over. When deleting words and pictures and memories becomes easier than saving it all, creating a line. It ends where we hopelessly decry lines, imagine fancifully the answer to be in circles, only to realize that fancifulness is its own trap, longing a contraindication in the very thing-that's-longed-for's-truthfulness. To want anything is to make it a god and the only thing that can be god is God—the rest give way to allusion. Like a longing for an afterlife, the more you want it, the more allusive it becomes. The way to get what you want is to stop wanting it. Just make sure by the time you get to the point of not wanting it you still want it, that there's still something left in you to want at all. Give me a circle, I'll give you a heart. Spin it. Just right. Spin it. Make it come true.

APRIL

ϟ

VENTILATORS AND ELECTRICITY

Graham Moore, the author of the screenplay for *The Turing Point*, takes us back to New York at the turn of the twentieth century in his novel *The Last Days of Night*, with Thomas Edison, George Westinghouse, and Nikola Tesla battling it out during the War of Currents. All three men are interested in science, but the similarities end there. Instead we have portraits of three very different men: Tesla the wily genius, Edison the astute businessman, Westinghouse the homely, grounded entrepreneur.

The fictional story revolves around the now-folklore of Edison having been the villain in this fight to light up the world. It's a pivotal moment in history, when science and business combined to make modern day technology, giving rise to the companies we know today: Westinghouse Electric Corporation, General Electric, and Tesla (which actually delivered ventilators to supply the shortage in Michigan).

The novel is told from the perspective of the young lawyer Paul Cravath, who Westinghouse hired for the Edison-Westinghouse light bulb patent lawsuit. Through Cravath's eyes the story takes an unprejudiced view of science and scientists, one that is neither (or both) divinized and demonized. Even as Westinghouse is fighting to switch from d/c to Tesla's a/c, Harold Brown is in cahoots with Edison in inventing the electric chair and petitioning the New York Legislature to consider it for the death penalty. The opening scene of the book is of a man being burnt alive while fixing d/c supplied wires.

We are presented with a view of scientists vying for power and prestige through all sorts of morally questionable behavior, adding to the notion that greatness and goodness are different entities, a theme touched upon in the work: becoming good—the higher ideal—often involves a misled foray into greatness.

The scientists we meet in Moore's work follow on the heels of the likes of Sir Humphry Davy, Michael Faraday, Zénobe-Théophile Gramme, and Paul Jablochkoff. Tesla by far is the most unique and brilliant, and his creative counterpart is found not in the other scientists but in the heroine, a singer by the name of Agnes Huntington—the sole female in an otherwise male-dominated narrative and a reminder of the dearth of women of science during this time period. We meet Alexander Graham Bell, inventor of the telephone, who mentions being in correspondence with a couple of brothers in Ohio trying to fly. We are introduced to a young Henry Ford, who is just starting off his career with visions of a grand business scheme. We are presented with both characters from prestigious backgrounds, like the Astors, and those who have fought their way through the world, including Edison, Cravath, and Tesla (who is a Serbian immigrant). Inventions of all kinds are scattered throughout the work: light bulbs, telephones, electrical wiring, generators, X-rays, motion pictures. The short chapters run like the scenes of a screenplay, making it easier for the production of the movie, with Eddie Redmayne rumored to play Paul Cravath.

What we are left with is a painting of Manhattan comparable to Raphael's *The School of Athens*, one of those spectacular moments when multiple geniuses converge in the same space and time, changing the fate of history. A Manhattan that is a counterpart to the Royal Society of Newton and Darwin on the other side of the Atlantic, this time American, international, and unashamedly proficuous.

As for changing the fate of history, the fictional Edison says with a nostalgic air that he sees science shifting from being knowable to unknowable: "It won't be like this. It will be more … technical. Inside the magic box, not outside it. A light bulb is intuitive; an X-ray is practically alchemy. The machines are becoming so infernally complicated that barely a soul can even conceptualize how they work. ... From here we can only build incrementally. Improvements. Not revolutions. No new colors, only new hues."

I think of imagination as the space between emotion and cognition, and that imagination gave a parthenogenetic birth to the twins of fact and fiction.

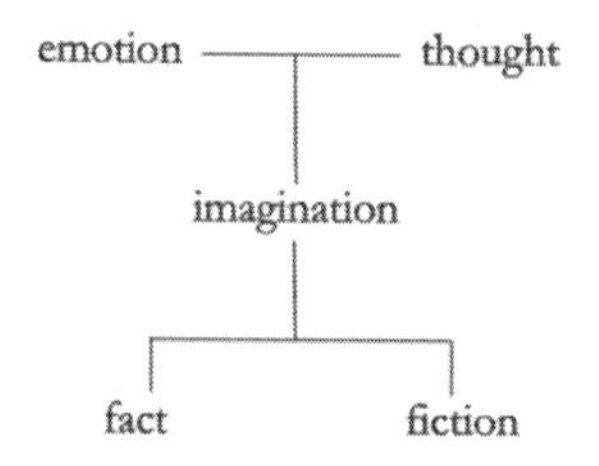

Approaching truth (if we allow it to exist) trapped between fact and fiction requires a commitment to goodness and a multimodal approach, and even then, this can only extend to the limits of what may possibly be a limited imagination. There's been a robust attack on the body of human knowledge as not having been "pluralistic" enough—or what I like to call the struggle against powered epistemology. The greats have been falling like dominos in popular imagination, a replacement in aching. From artists like Picasso, whose African period has been under hearty critique, and David Foster Wallace, whose questionable relationships with women came to the fore thanks to the #MeToo movement, to the innumerable scientists involved in racist eugenics. If we don't separate knowledge and art from the idea of power and their creators, we are left on very tenuous bearings. If we do, we are still on tenuous bearings. The only concession left is to attempt to undo the individuals involved in the evolution of art and science in the last few centuries.

I'm going to get a little more elaborate with this thought experiment and dot out and bend the tree I drew earlier:

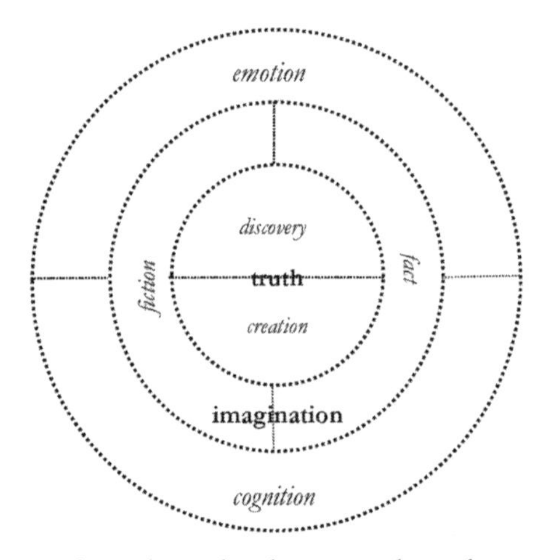

Scientists have been harder to replace than artists. The Sims retractor is still in use in delivery rooms around the world, and even if it were to be renamed, the object itself cannot be taken out of the world and history. Academia is attempting to replace the eponymous discoveries or inventions that are strewn across the scientific world. Rosalind Franklin's re-addition to the scientific liturgy, James Watson's stripping of his Nobel prize, the Dev Patel movie on Srinivasa Ramanujan's life, Netflix's documentary on Abdus Salam are examples of a mainstream effort. What I assume the attempt to be is that of a *School of Athens* presenting a diversity panel in the way of a Maybelline ad with the faces blurred out.

Art is superficially more expendable in that vein. Remove a painting from its central position in a central hearth of a central museum in the center of a Western metropolis and the object has been removed. In science you can take away the name but not the object, whereas in art you can take away the object but not the name.

There have been steady openings in the heart of art and Frida Kahlo has been often used to fill them. Her paintings were exhibited in the Detroit Institute of Art earlier this year. At one of our intern orientation lectures we were shown a slide of Kahlo's *Henry Ford Hospital.* It depicts her miscarriage at our hospital, the artist prone at the center of a hospital bed in a pool of blood, with flying objects tethered to her ripping body. The six objects tied to the ends of the ropes include a uterus, a flower, an orchid, a snail, a pelvic bone, and an autoclave.

The autoclave is a device used to sterilize instruments. It uses pressurized steam reaching above 250 degrees Fahrenheit for about twenty minutes. It already existed before Tesla and Edison's time, hardly anything alchemic about it. A present-day ICU room similar to the one in which Frida Kahlo had her miscarriage would probably be stocked with a ventilator more akin to a magic box.

There was recently an article in *Time* about the history of ventilators, a commodity at the center of attention given its need for covid patients with respiratory failure. The "iron lung" that was developed in the 1920s for patients suffering the complications of polio relied on negative pressure ventilation. The first modern medical respirator that relied on positive pressure ventilation, the Mark 7, was developed in the 1950s by a former World War I U.S. Army pilot.

Art has always been open to an inexplicable and inexhaustible data source. The same cannot be said of science. The world was somewhat explicable until it wasn't, right around the 1920s. If Diego Rivera had come to visit Frida at the hospital today, he probably would have had a laptop and cell phone on him, the television would be on in the background. We live and function in a technological world which I don't think anyone understands even crudely how it all works. It's almost enough to paralyze wonder altogether. But a way to disable that paralysis is to un-individualize the self when it comes to science: to believe in knowledge greater than the individual.

What's the purpose of knowledge that isn't timeless or reaching for itself though? All that effort only to either end with death or dementia or the end of human existence at some point, whether we evolve into another species or just end full stop. If it's possible to create meaning outside of temporality it would be possible to justify its purpose—many writers and artists do a beautiful time of it, beauty as a source of meaning a vivid alternative to time. Meaning is too intricately connected to time for me though to do without the soul and its blue speckled hues.

☺☺

PLUS, ONE

Time out

It's the evening of Wednesday, April 1, and I have my first day off tomorrow in more than two weeks. The Holiday Inn I'm staying at is usually full but currently has at most seven guests. I've actually encountered only one other human here: the bellboy. I'd forgotten my room key and went to ask him for a replacement, when I noticed him painting at one of the lobby tables. I asked him what he was working on and was shown an abstract of two faces, not quite the comedy and tragedy symbol of Sock and Buskin but I'm going to pretend it is.

I put on my parka and went on a drive to get coffee from Tim Hortons and sat in the parking lot going through my social media, where the sole topic of conversation is this pandemic. Emotions can either be refined or dramatized, and there's a pretty bold line between the two. It is possible to approach the dramatic without turning it into a drama. I've seen posts about being out on the front lines that have literally, physically made me cringe: an overdramatization that's become a caricature. Without the levity afforded by humor, situations and humans sink into farce. There are conditions where humor in catastrophic situations can become grotesque, like when we had to terminally wean an elderly woman off of oxygen yesterday because she'd declared her wishes prior to arrival as DNAR (do not resuscitate). However, I don't think that that applies to the entirety of the current circumstance. Humor, like the frequent reappearance of Monty Python,

Key and Peele, and ZDoggMD in our workroom, instead of sublimity, has been on the whole easier to entertain.

Plus, one

I Facetimed with my best friends, one in New York and one in Lahore, later that afternoon. We talked about how the supposed third world's initial reaction has been less absurd than the first world's. Part of it may be due to the fact that the third world has been used to the world ending. The world has ended for us many times, so we live in a baseline normalcy of temporality and unpredictability. There's a dependence on and respect for seasons and the world's whims instead of acting as if independent from the earth, which is what it sometimes feels like on this side of the world.

I wouldn't want to be too exacting though. It is remarkable what that opposition has brought, including the fact that three different spaces could be united when we were Facetiming. Things sometimes swing back and forth before balancing out—kind of like how the world swung from phone calls to text messaging to now balancing out into a youth that mostly communicates through voice notes and disappearing selfies. The written word is great contractually and for performance, but aural and visual communication are the media of connection and growth.

Before this pandemic I had been sure Trump was going to be re-elected, but I think that's questionable now. America's response to the pandemic has been criticized nationally and internationally. There are pictures juxtaposing American healthcare workers in embarrassingly basic PPE with their better-equipped counterparts in Italy and China. There are articles on the resistance to wearing masks and social distancing and the economic restrictions that would necessarily entail.

I drove home later in the afternoon to pick up some food. I kept my windows rolled up with my N95 mask squeezed in place, while my parents waved from the garage and put out a large suitcase of food for me. They closed the garage door before I got out to put it in my backseat. I'm a pretty emotional human, but wasn't expecting the constriction of my throat as I backed out of the driveway.

When we die

I returned to the hospital the next day to learn that a nurse on another floor had died of the virus. She's dead and we went on and are here. Turn that "she" into many and that's where we stand today. How do you encounter the depth of something without dramatizing it? Keep it simple, less wordy, a gesture—a bowed head.

My first few encounters in the autopsy room in my first year of medical school had left me devastated, and in the guise of formalin irritation Lolita's something, something, those something nights, and the cars, and the bars, and the bars, and the barmen would run through my mind as a kind of chant. I went to the cafeteria one day to get a cup of chai and found myself next to a sickly, emaciated man who looked so much like one of the cadavers I'd been harnessing that I smelt the numbing formalin.

I made it through anatomy into our third year, and our curriculum that year involved forensics. So many images, so many brutally killed, charred, raped, sodomized. From an intricate, infinite creature, spent years learning, details of anatomy and physiology, reaching not even a fraction of reality, to end like this? How to live life, constantly on the brink of this? Where were their souls? All my eyes impressed upon me was death. Death, from contact flattening, to changes within the first twelve hours—pallor mortis, algor mortis, rigor mortis, livor mortis (cooling, pooling of blood, stiffening, postmortem staining), to eventual putrefaction.

How could a soul leave if death was not a moment, but a process? It somehow more easily made sense if it were a moment, a snapping, in one way, out the other. But a process made it an entirely different question—a process divided into somatic and molecular death, a process whose boundaries are being pushed further and further by medical advances and the reason organ transplantation is so successful. Was it the gradual disentangling of a beautiful soul, weighing the twenty-one grams that Dr. Duncan MacDougall's flawed experiment calculated it to be, that left the sham of a body in ugly disfigurement, or something far less fanciful? Good thing immortality (religious or otherwise) goes hand in hand with death, otherwise maybe that would really be the end of us. Allow the connection, or at least be aware that it exists inside of you where it does, and move on.

Novelty

There's been a new protocol almost on the hourly. I returned after my first day off already feeling as if I'd missed out a significant part of the plotline. Duck-face mask in place, I sanitized my phone, badge, keyboard, and mouse with bleach wipes before taking a seat at my desk. The medicine chairman and infectious disease specialist made their normal daily morning stop-by, asking us how things were going, questioning new things we were discovering, and providing updates. I opened a patient's CT chest read and the radiologist had quoted the "Radiological Society of North American Expert Consensus Statement on Reporting Chest CT Findings Related to COVID 19." It had been published on March 25, by Scott Simpson et al., and was already making it onto radiological reports.

The first few days on the covid floor we had gone into our patient's rooms with our attendings, but that's no longer how we're doing things. The staff goes geared up alone into the patients' rooms, while we stand outside and catch the outgoing face shields and stethoscopes that need wiping down with bleach water in pale pink plastic buckets. This minimizes the number of people exposed and saves PPE.

Before covid, once a patient started becoming hypoxic on 6L of supplemental oxygen through a nasal cannula we would start thinking about possible transfer to the ICU. However, despite the addition of ICUs in the hospital, there haven't been enough beds to which to transfer patients. As an alternative, we're starting to flip people onto their bellies once they start requiring more than 6L of oxygen, to allow greater lung oxygenation. Some of the staff cutely refer to it as "tummy time." It causes further anxiety in nursing staff who can't be in and out of the room as often as needed to make sure their patients aren't suffocating due to safety precautions and having to gown in and out.

On that note, here's another technical update:

1) Azithromycin fell though as being a treatment option. Azithromycin and hydroxychloroquine both prolong what is known as the QTc, the interval in the EKG that correlates to the depolarization and repolarization of the heart's ventricles. This increases the risk of going into

cardiac arrhythmias and has already caused deaths in the country. Instead, we're only using hydroxychloroquine with regular EKG s to monitor the QTc.

2) We're starting to consider getting CT chests in patients whose D-dimer continues rising while the rest of their covid labs start downtrending. We know that the cytokine storm increases hypercoagulability and therefore increases the risk of clots. D-dimer is actually used as a surrogate marker of increased coagulability and one of the tests we get to sometimes determine the probability of a pulmonary embolism. Do you remember me mentioning earlier that a study had shown increased D-dimer in people who had died of covid? It is possible that those people had actually died of pulmonary emboli as a result of hypercoagulable states. We started religiously prophylaxing our patients with anticoagulants, and despite that still have incidents of patients developing pulmonary emboli.

3) The virus uses its spike protein to bind to ACE-2 receptors on cells. There have been conflicting studies whether to continue or discontinue ACE inhibitors, a class of antihypertensives, in covid-positive patients. Although the most recent guidelines said to continue them, if the sole reason for patients being on the medication is hypertension, we are opting to temporarily control their blood pressures with different classes of drugs.

4) Our hospital is starting to use plasma exchange and convalescent plasma treatment in select patients.

Something, something, something

The collective nature of this experience (#alonetogether) has generated many metaphors, including the military language that's been used to describe it: "redeployments," "victims," "frontlines," "martyrs." There's a picture in our living room of my maternal grandmother with her medical year cohort standing outside the entrance to King Edward Medical University in Lahore with Eleanor Roosevelt, a garland around her neck and a bouquet in her hand, in the center. My grandmother was in medical school when the British decided to bounce and Pakistan and India were partitioned. Approximately one million people died in one of the bloodiest redefinition of boundaries known to mankind. She was redeployed to sift the alive from the dead in the trainfuls of humanity arriving at her end in Lahore. She recounts those stories with an enviable amount of equanimity and faith. It happened, it was devastating, it was, it is.

Some of us are over everything and ready to go back to our old lives while others are excited about potential changes. Our neurology department drew up a protocol for various neurological presentations for the ED and are hoping that a part of that protocol will be continued even when things settle down. The first encounter with this pandemic was fraught with fear and gusto, and the initial confusion ironed out into a semblance of intelligibility and eventual boredom for some.

We are outside the realm of strict guidelines and answers. Instead, we're in the realm of creative medicine, and that part of it has been thrilling. My final year medical examination comprised roughly five months of written, oral, and practical examinations. We would stay in the hospital studying patient cases and fine-tuning our clinical examinations from morning to night. I had tried tucking death away with her annoying questions, somatic and molecular, but she'd beat fists against her compartment: while CPR was being performed on a dying man; while the women down the hall started wailing, indicating the end of her loved one's life; while two elderly patients died on the same hospital ER bed because of a lack of resources. The longer I stayed in the hospital, the more persistent she became, contact flattening, a pressure trying to make her way through the barrier.

She was finally released in states of mental and physical exhaustion, after having seen the worst and best of humanity and entering a kind of sacred plane, laughing at the end of the day with the people I worked with. If you're able to somewhat reconcile (if not necessarily unify) life and death, you feel a lightness. In the middle of those final exams I was surprised to find moments where suddenly everything fell into place, click, into my palms as if it all was and would be and this is all there was and would be and is.

That was the first time I started looking at medicine not as a walkway between life and death, but rather as a creative process of healing and innovation. Looking at medicine through that lens, the questions change, emerging from creative discovery.

THE CASE OF THE MISSING FEATHER

We pointed to the drawing: "What is this?"

"A feather," Mr. Quincy pronounced, and immediately started crying, to our astonishment.

"Why are you crying Mr. Q?"

"Because feathers make me sad."

"Why do they make you sad?"

"Because they remind me of Native Americans." Tears broke free of his lower palpebral ridge.

"Why do Native Americans make you sad?"

"Because we killed them" he sadly sniffed.

"Were you there when they were killed?"

At this point Mr. Q stopped crying and grew pensive: "I'm not sure."

Mr. Quincy had been an overnight weekend pickup for me. At a large tertiary center like ours, weekends can be infamously busy, with patients from around the greater area being transferred with everything and anything due to local facilities calling it a weekend. Mr. Q was an older gentleman who, according to the transfer system, was presenting with altered mental status suggestive of delirium. The outside facility had worked him up, then sent him to us. The workup at the other hospital had included a lumbar puncture that had mistakenly come back positive for bacteria in the cerebral spinal fluid. He had been started on an antibiotic called cefepime for possible meningitis, before the labs came back as a false positive and he was taken off of it. Looking at the chart, the senior resident and I were assuming he had come in with some kind of infection, maybe a garden variety urinary tract infection, that had caused delirium and had been exacerbated by cefepime, an antibiotic known to cause altered mentation.

But as I started examining him felt something incongruent. This patient didn't seem delirious. (Delirium is differentiated by dementia by its waxing and waning course and usual temporary duration.) His arms were tremulous at baseline and his limbs had been hypertonic. He had cogwheel rigidity. His reflexes were all brisk. No overt signs of a focal deficit. When we came back as a group to examine him during Monday morning rounds, our exchange about the feather left us further puzzled. He started crying again at another tangential association. He could not recall who he lived with and was unsure whether he was married or not. This had to be some kind of rapidly progressing dementia. We started coming up with differentials as we left the room. Prion disease? Pick disease? I went back into his chart and saw what the previous physician a week ago had noted: Lewy body dementia.

Lewy body dementia is characterized by Lewy bodies, aggregates of alpha synuclein proteins in nerve cells, that are also found in Parkinsonism. The differential fit. The family had described a characteristic stooping posture and shuffling gait that had developed over a year or so.

My co-intern started discussing the theory that Parkinson's disease begins in the gut (strange how it always seems to come back to that). The name for the vagus nerve, the tenth cranial nerve, comes from the word "wandering," for its long course and supply of laryngeal and pharyngeal structures and the thoracic and abdominal viscera. The theory is that alpha synuclein bodies, aka Lewy bodies, are formed in vagus nerve ends that supply the abdomen and travel retrogradely into the vagus nuclei in the brainstem. Patients who have undergone vagotomies have been further studied to support this claim—if your vagus nerves have been cut, so has the question of retrograde transmission into the brain. The question still remains why and how Lewy bodies form in the gut in the first place.

A few rooms down from Mr. Quincy's into our morning rounds and my co-resident and I start debating consciousness, whether it was even a useful term. To say something is "truly red" and that it is "red" are the same thing—"truth" is redundant. Can the same be said of consciousness? To speak of the mind and consciousness and the brain and the soul, it all seems to have become synonymous and outmoded (though I'd emotionally fight myself blue that the soul was always distinct, by definition elsewhere—a younger me asking my father if he'd ever seen the soul while holding a brain, and if that made a difference). Neurologically, consciousness is described as the awareness of the self and one's surroundings. What's your name? Do you know where you are? What's the year? Do you know what's going on? We often only know things by their loss, and there's a sequence to loss of memory: Time is the first to go, followed by place, then person slips out silently through the back door. Mr. Quincy was oriented to self only. When I asked him if he knew the date, he replied, "It depends on the nature of the sun in its compensatory orbit around the earth." I didn't correct his pre-Copernican geocentrism, but I found it an interesting answer nonetheless.

The next morning, I went in to perform the dementia scoring test, MOCA, and Mr. Q brought up how he loved letters and numbers. How every letter had a number and they were all codes that had to be made meaning of. I excitedly remembered *A Beautiful Mind* and was comically disappointed when I asked him to give me an example. He spelled out S-P-E-R-M. I asked him what the hidden meaning behind that was. He said it meant "equal rights for men." There were no numbers involved, and it no longer mattered what had been said or could have been said. He started talking about how the tubing at the end of the bed was trying to talk to him. He'd say things and associations that disappeared within seconds. When I asked him to draw a clock, he started drawing a cube. I pointed that out to him only to be told, "Well, it could be a square clock." His clock was squiggled into one face of the cube. Fascinatingly, there's a whole literature of dementia pathologies with drawing of clocks that would do Salvador Dali and *The Persistence of Memory* proud.

Treating Mr. Q was difficult. He would sundown and become extremely paranoid at night, screaming bloody murder. Dopamine is the neurotransmitter involved in both psychosis and Parkinsonism, which makes treating both conditions coming together in Lewy body dementia difficult. If you try to treat psychosis by bringing down dopamine levels, you inadvertently make the movement problems worse, and vice versa. We couldn't give him antipsychotics, so instead of targeting the dopamine levels in Lewy body dementia, we focused on another neurotransmitter, acetylcholine, which is a target in Alzheimer's.

A few days later, close to discharge: "It shouldn't upset me this much, the comings and goings of reality." You and I both, sir. Mr. Q was having ideas of reference: besides the tubing, any kind of plastic was trying to reach out to him. He was always surrounded by well-formed hallucinations, of humans and animals alike. A dark-haired Italian woman in the corner, squirrels jumping all around him, my phone constantly ringing to his ears when it was on silent.

In *The Lost Mariner*, Oliver Sacks talks about a man named Jim with Korsakoff syndrome, memory loss caused by alcoholism. Sacks asks the nuns if they think Jim still has a soul, and the nuns angrily tell him to come see the patient at chapel. Sacks goes to see him and writes the following, "I did, and I was moved, profoundly moved and impressed, because I saw here an intensity and steadiness of attention and concentration that I had never seen before in him or conceived him capable of. … Seeing Jim in the chapel opened my eyes to other realms where the soul is called on, and held, and stilled."

That passage by Oliver Sacks reminded me of *Mémorable*, a 2020 Oscar-nominated animated short film, that depicts a painter's progressive decline into dementia. The neurologist in this short film is literally and physically depicted as a one-dimensional character, in stark contrast to the painter's wife. There is a magnificent moment towards the end when the painter, who has reached severe dementia by that point, dances with his wife and they turn into a twirling unit of three-dimensional colors.

It's hard enough to label the soul when it's still, an impossibility when it becomes as labile as it does in dementia. Instead of searching for a fixed silhouette, the attempt with dementia becomes a silhouetting of a dynamic waltz. Futile from the get-go. Again, we learn of things by their absence, and I think that's why we look towards dementia when we think of the soul, but I'm not sure that's always reasonable. First, that makes consciousness and the soul synonymous, which I don't think they are. We'll let the definition of consciousness remain the neurological one, but the soul is defined as the immortal part of a human. The definition of immortality can be up for grabs, whether you think you want to be immortalized as the stuff of stardust or a more concrete spirit-stuff or an actual bodily resurrection.

I've seen enough to limit intellectualism in the face of suffering. The way it breathes away from everything you think is real until it's not. Until you realize that it falls short in the face of humanity. I've seen a woman lose herself to the grief of her husband being eaten away by jaw cancer. A man in faceless loss by his wife's end-stage liver cirrhosis. Mr. Q's wife facing Mr. Q, no longer really her husband. It's not because of suffering that there must be meaning; it's the idea that meaning could be either limited to or scalloped away from it that's absurd. The idea that suffering is meaningless is meaningless, because what is the point of reference for that claim? A simple statement of having been heard, whether in silence or anguish, suffices. Both torment and peace, heard. At least that's what I'd hope for.

"Do you see a river flowing in that chair?" Mr. Q asked the physical therapist working with him. Mr. Q was discharged home only to make it back to the hospital a few weeks later, at the start of the pandemic, to be discharged quickly from the ED. Brains can be magnificently poetic when broken. The tragedy is that that brokenness comes at the cost of an insurmountable loss. I'm no longer in the business of seeking a soul where it can't be found, as alluring as that trap is. I can still appreciate the science of pathology and fall in love with the manifestation of pathology without getting caught up in the philosophy of pathology.

THE THINGS WE LEAVE BEHIND

It's Thursday, April 8, and I had to take the day off because I was unwell. My covid swab came back negative and I was left with a day off during which I watched the Joan Didion documentary *The Center Will Not Hold.* Months earlier, my friend had sent me one of Didion's articles, her famous one on self-respect, published in *Vogue*, and I had fallen in love. Her assertion that self-respect is at the basis of both love and indifference make more sense to me than the love all preachments.

One of the most annoying things about living in a hotel room is the lack of air. There are no windows to crack open and it leaves the mechanically heated room feeling unforgiving. However, the airtight stillness is an appropriate environment for the first half of *The Center Will Not Hold*, which was so geographically rooted in California, with a brief stint in New York, as to be almost unrelatable. The story of a mind unraveling in New York is common enough that in my NYU freshman writing class we were told to stay away from the topic altogether. Of course, this was Didion, so she's allowed to break that rule—or maybe she predated that rule. She embodied her time and generation through essays. What of the opposite though, those who cannot embody time or space and can only mirror back a sense of displacement?

Most of my close friends aren't physicians or even in the same time zone as me, something I'm mostly thankful for, but that also means I've missed a lot of important events: wedding showers, weddings, baby showers, babies being born, first and second birthdays of aforementioned babies.

Storm of babies aside, even blood needs blood: vasa vasorum. You need external feeders of similar internal force. You are who you're willing to let go of—don't take aphorisms away from me, instead let's dip them in sugar, feed on them on Fourth of July driveways, blue jeans shredding the tar like chalk. Meaninglessness was different for me: how I carried it away.

Didion talks about the meaninglessness of experience in LA, and I thought: 1) a large part of her meaninglessness comes from a cultural rootedness and 2) that's definitely not specific to LA. Meaninglessness is experienced differently in different places. In Lahore it was light and temporary and red. In Michigan I've experienced it as something deep and unforgiving. In New York it had been a brilliant nothingness.

The second half of the Didion documentary changes tone, after she becomes involved in covering politics and lives through the deaths of her husband and daughter in quick succession. Her daughter, who suffered from alcoholism, was rushed to the hospital. The documentary homes in on the dark entrance to an emergency department sporadically illuminated by the red lights of the ambulance. Didion's daughter passed away eighteen months later, due to various complications; the rest of the documentary pivots to her memoir *The Year of Magical Thinking*, about the abrupt death of her husband, John Gregory Dunne, and its aftermath.

People are dying not just because of covid but because the system has shifted, the holes in the Swiss cheese model are aligning. Anything other than strictly covid has become a question mark, and alongside the covid deaths are the technically non-covid deaths directly related to this pandemic effectively shutting down everything else.

I returned to work the day after the Didion documentary to my first covid-positive death. In fact, to my first death. It was the first patient since I'd started my intern year to die in front of me. It was my first call to a family about their loved one having died: Can we talk for a bit? Are you sitting down? I'm afraid I have some bad news. And, once again, it wasn't poetic and was a few bends traumatic.

Despite the predominant cultural theory that physicians must exhibit emotional exactitude, I was hesitant to emotionally distance myself for a long time—to give in to what I considered apathy in med school. When I started residency, I'd tear up with patients and family members. Then (surprise) I realized that frayed nerves are a drawback. That much emotion in a day jam-packed with patients would be unbearable for any set of nerves (poor Mrs. Bennet, I'd never imagined I'd come to sympathize with her). And even if it could be tolerated, there are enough articles and studies that show, fortunately or not, that the best way to deliver care is to detach ourselves. Physicians are supposed to be emotionally distant certainties in uncertain circumstances. The reason it's not a good idea to treat your own family members is because of the high emotional investment. You need to know when and when not to envision patients as family members.

Are you sitting down? I'm afraid I have some bad news. I know she's not mine, and I can't quite be yours, but I swear some part of this, even if it's outside of me, still matters. I wonder if the voice on the other line felt like his world had changed. I don't believe in moments, the idea that change can happen with one event or occurrence. There are trajectories punctuated by experiences, albeit significant experiences. But there are no moments; there are only changes that we wish were moments. Or maybe they do exist, mostly in the form of loss, and I'm being obstinate here. You've only known loss if you can point to a moment (that winter they shut the gas and we had no heat). You've only known loss if you can stretch it across albums (that winter they shut the gas and we had no heat). Are you sitting down? I'm afraid I have some bad news.

I returned to the workroom to document "the death note"—words used to describe death justified over death itself. After work, the cold Detroit air rippled against the edges of hysteria as we walked to my co-intern's resident apartment across the hospital. We scalped through what had happened, what should have happened, what could have happened, and talked about natural phenomena and life after death, in which some of us believed and some of us didn't, and I realized the futility of saying what is what. What if, what if, what if.

This was a test of feeling without wanting: the stage set for "The Physician Grieving Her Lost Patient." The necessity for solemnity undermines itself. It felt blasphemous to laugh but at the same time somehow ridiculous to put on an air of solemn purpose. I remembered the autopsy room in med school and the crassness of other students, sneaking in food and laughing amidst the dead. Perhaps immortality and death were so intertwined for them that they felt no need to respect something that they felt was no longer there, a body without a soul, and only a soul worth respect. "The Physician Grieving her Lost Patient" would depict a scene beyond crassness but with tethered homeliness: here in the plains, we know a cascade when we see one, here in the heated dust, we know an ache, stuffed into soft cut dough, when we taste it.

Cupped full of Buddy's pizza and coffee and unwilled laughter I returned to my hotel room that evening and called my parents. The image of her dead body still blinking. Stones of the eyes; stones of the heart. My mother recounted a story from her house job in one of the public hospitals in Lahore. There had been a young guy who had come back from Saudi Arabia after having made some money. He'd gotten into a street fight with a former friend who had been jealous and stabbed him. They had brought him to the hospital where after the stab wound was controlled, he had gone into acute respiratory distress. She said at that time there hadn't even been one ventilator in the hospital, and they had watched as this young guy died. You never forget.

MAY

MORTICIAN

The hospital library is on the seventeenth floor and the mortuary blessedly somewhere below. I was with a few medical students and we were on our way to the ground floor when a mortician stepped onto the elevator with us.

"Oh, it's the doctors."

Silence, some shuffling bodies, some hellos.

"You guys just end up killing people. All day I see the end of your screw ups."

Awkward silence.

"Look at my finger here," she pronounced while holding up her index finger, "I had trigger finger, and all these people including my son wanted me to get surgery on it, and you know how I fixed it? I drank some herbal tea and I never needed surgery. You know what would have happened if I had gotten surgery?"

More silence. Some nervous nodding, people looking at corners of the elevator waiting for it to reach the ground floor.

"Well, let me put it like this. I get the products of what doctors do."

We reached the ground floor and burst out laughing once the mortician was out of sight.

This woman sees a skewed population sample and has formed an accurate opinion based on available data. There are as many books about the power of medicine as there are on the failure of modern medicine. Most are read by very different populations that rarely overlap. Humans tend to search for data that validates their own beliefs instead of making the uncomfortable journey into a different dataset, something that's known as "confirmation bias." It's safer to stay in one circle. Once you start trying to step outside of it there's no way back in, no bigger circle encompassing you, it just becomes a weird tortuous spiral. The mortician's regular supply of dead bodies is enough for her—a waxed arm, a waxed leg, a waxed torso enough to form a belief.

It's weird how movies about dead bodies look more realistic and the more dead bodies you see the more you think you're seeing the representation of death instead of death itself. I'd seen a bomb-torn body—arms, legs, torso—on a silver tray in the forensics lab in my medical school through the lens of a movie instead of reality. It's even stranger that Socrates (or, well, maybe just Plato) was talking about that even before the advent of pictures and cameras and today's technology. It's similar to unmooring meaning from words by repeating them.

The longer and more frequently you see something the more it disconnects from the ideal it represents until the ideal takes precedence. I visited my friend's village outside of Lahore one summer and slept on charpais under a night sky so surreal that the only way it felt real was to call it the Great Hall in Hogwarts, which in the series features a ceiling that mimics the weather outside: The Great Imitator. There are a few systemic medical diseases that go by that label. It used to be used for syphilis before the age of penicillin and is now used for lupus, sarcoidosis, and Lyme disease, to name a few.

Death can also be The Great Imitator. Not in the sense that it's a difficult diagnosis, but in the sense that a dead body only imitates what we think of when someone says "death." Death is the cessation of life, and if we take away the metaphysical implications of life that's a pretty obvious thing to point out. But that's the thing, we define death by what it's not: it's not life. We know life intimately before we know death, and their temporal relationship makes it hard to see death without life in the equation. It's easier to imagine what life without death would look like, it's much harder to imagine what death without life would look like. The sequence ruins everything else. Or maybe I need to cognitively reframe that: it doesn't ruin everything, but it certainly impairs something.

What would knowing death before knowing life look like? The closest approximation is writing this now as opposed to when I couldn't get the dead covid patient and her mind out of my own mind last night in my hotel bed—a body, a body, djinns in the hotel cupboard cackling at the way we lick fire with fire thinking we're living on the undersurface, but in reality we're playing with the surface, twirling above it below it within it, if you will not bend you will be made to bend, upon my bent, cross my heart, teach me how to soften.

After unmooring the meaning of death from the image of death, you turn back to the reality of it. Ultimately, we conquer death by turning away from the ideal towards reality—hope ironically follows.

HOLES IN THE BRAIN

Today—where are you. Tomorrow—where are you. Yesterday—where are you. I'm in bed in my family home's basement, where I'm now isolated after moving out of the hotel, and trying to recreate my story, consciousness just the lateralization of drowsy early morning thoughts extending to the ears and dangling down like chandelier's—Beyonce's super power. Clad in a pink power suit and chandelier earrings, her and Jay-Z stood in front of the *Mona Lisa* in the Louvre for the music video of "Apesh**t." The video itself is a beautiful critique on the history of powered epistemology.

The history of medicine may have begun (we like beginnings and firsts) with the history of trepanation—burr holes made into the skull to access brains overtaken by spirits. Now, it's the twenty-first century and burr holes are still a procedure with, granted some differences since its inception, including the indications (spirits replaced by brain swelling), the sterile surroundings, and the fancy equipment. I think it's sometimes important to look at history to see how far we haven't come. It's good for human humility. The curse of history is that seemingly wondrous present-day discoveries are seen with suspicion—as something that will only inevitably become barbaric in retrospect.

The first archeological evidence of trepanation is reported to date back as far as the fifth millennium BCE. Holes were bored into skulls in an attempt to purge spirits that we can only assume were in actuality some kind of mental illness. Some of these early humans survived and some didn't. The process of mummification in ancient Egypt involved removing the brain and rinsing the skull with resin and wine. The heart would be left in peace. We have evidence of Egyptian medicine left on papyrus remnants, where most disease was considered to be caused by bad food—also not completely in contraindication with how we think today.

Health temples devoted to Imhotep in Egypt were transported to temples of Asclepius in Greece. These temples were often inhabited by snakes that kept temples clean of rodents and whose molting served as a symbol of health. The Rod of Asclepius is the symbol we associate with medicine today: the snake wrapped around a rod.

The father of medicine is famously known as Hippocrates (460–370 BC). He started separating medicine from the spirit world and postulated the function of the four humors—black bile, yellow bile, phlegm, and blood—in health regulation. However, bleeding, purging, enemas, and massages as practiced by the Egyptians were continued.

Next in stature to Hippocrates was Galen of Pergamum (129–216 AD). He was a personal physician to the Roman emperor Marcus Aurelius, the last of the Pax Romana. Galen tied motor activity to nerves but endorsed Hippocrates's humoral pathology, which would last for thousands of years.

While the Roman Empire rose and fell, the Islamic world made advances in sciences, specifically pharmaceuticals, physics, chemistry, and astrology. Some famous names include ibn Bultan, ibn Khaldun, ibn al-Haytham, Avicenna, and al-Razi. Peter Adamson's book, *Health: A History*, talks about al-Razi's claim to fame having been the use of control groups in medical experimentation, paving the way for modern day science. Then, of course, there was Avicenna (980–1037), whose *Canon of Medicine*, a fourteen-volume work, was used worldwide until at least the eighteenth century.

According to Allama Iqbal, empiricism was built into the Qur'an, and for that reason there was neither a significant nor a lasting debate between the real and ideal in Islam. The translations of Greek texts became common during the Abbasid dynasty, and although Greek thought broadened philosophy, for almost two hundred years it obscured the anti-classical nature of the Qur'an. There is a reason then that Islamic thought could never form a basis in Platonic thought and was more in accordance with Aristotelian and Peripatetic schools, until eventually with ibn Khaldun the anti-classical and empirical basis of the Qur'an was set.

The largely non-narrative form (spots of stories in a background of what seems like repetitive chaos) and the names of the 114 Surahs (chapters) of the Qur'an assert its foundation in nature: The Bee, The Cow, The Sun, The Moon, The Cattle, The Thunder, The Light, The Ant, The Spider, The Smoke, The Star. One of the most commonly repeated verses in the Qur'an details the signs seen in the alternations of the night and day. The known world is given priority over miracles in the scripture.

Islamic thought was anti-classical in that it did not rely solely on theory like classical Greek philosophy. According to Iqbal, it was its empirical nature that formed the scientific method as we know it and allowed science to flourish in the Islamic world and beyond. It is empirical not as a textbook, but in that the Qur'an was used to foster an empirical way of thinking, of looking at the world. For example, rather than being metaphysical, individual immortality, claims Iqbal, can be based in biology, that is, in evolution. He writes:

It is strange how the same idea affects different cultures differently. The formulation of the theory of evolution in the world of Islam brought into being Rumi's tremendous enthusiasm for the biological future of man...On the other hand, the formulation of the same view of evolution with far greater precision in Europe has led to the belief that there appears to be no scientific basis for the idea that the present rich complexity of human endowment will ever be materially exceeded...Nietzche's enthusiasm for the future of man ended in the doctrine of eternal recurrence—perhaps the most hopeless idea of immortality ever formed by man.

I don't know if eternal recurrence is the most hopeless idea of immortality ever formed by man; Iqbal missed out on the whole "we're the stuff of stardust" fad (though that may be more inane than hopeless). I think there's much in it that can be similarly found in other doctrines. Embracing suffering to the point where you'd live through the same life willingly again and again does not sound particularly tragic to me and actually kind of reminds me of what you try to do in Sufism anyway. Anyhow, with the Mongols came the tearing down of the famous House of Wisdom of Baghdad that had been built during the Abbasid Caliphate and it's time to get a coffee and move elsewhere.

We move back into Europe, where Pope Sixtus IV (1414-1484) allows executed criminals to be dissected, marking a paradigm shift in medicine. Physicians and artists take to the task, amongst them Leonardo da Vinci and Michelangelo. Fascinatingly, in 1990 Dr Frank Meshberger published a paper in the *Journal of American Medical Association* on how God as depicted in Michelangelo's *The Creation of Adam* was an anatomical cross section of a brain. Once you see it you can't unsee it.

The formal father of anatomy, because of course there must be creators and fathers, was Andreas Vesalius (1514-

1564). We're towards the back end of the Renaissance and the beginning of the Enlightenment and the metronome is about to quicken. These early, relatively slower, millennia were a working out mostly of the physicality of the body. When it came to pathology, different theories such as vague contagion, divine decrement, bad diets, miasma, and humors had floated around for basically most of our existence. The next few millennia would expand upon the pathology and other branches of medicine. Glossing over the distinction between health and disease, the causative agents of disease have narrowed down in the history of medicine from spirits to humors to contagion and organs to tissues to cells to genes.

Giovanni Battista Morgagni of Italy (1682–1771) was the beginning of the end of humoral pathology and a turn towards the identification of organs as the culprit in disease states. With the French Revolution came French pathologists Marie-François Xavier Bichat and Jean Cruveilhier, who further honed in on organs and looked at the tissues that comprise organs. Cellular pathology would later take on its form with names like Schwann, Henle, and Rudolf Virchow.

Ignaz Semmelweiss famously discovered a correlation between physicians in a Vienna hospital with poor hand hygiene and puerperal fever in laboring mothers, however his findings weren't popularized. He developed some kind of dementia later in his life and died in a mental institute. Louis Pasteur was the one who would make germ theory a force to be reckoned with. Robert Koch would follow on his heels and become famous for linking certain diseases to certain organisms.

The history of surgery can be narrated as a parallel tale of surgeons that were stigmatized by the rest of medicine and considered on par with butchers. The story converges with the rest of medicine with anesthesia, Dr. John Snow chloroforming

Queen Victoria during delivery, and the evolution of antiseptic, and eventually aseptic, surgical procedures.

Molecular biology and genetics revolutionized medicine into its contemporaneous form. In 1940 Americans Edward Tatum and George Beedle demonstrated that there was a connection between heritable genes and protein structures. Oswald Avery linked genes to DNA in 1944. X-ray diffraction data collected in the 1950s by Rosalind Franklin was taken (stolen?) by Watson and Crick to formulate the double helix model (cue rescinded Nobel prize). Cystic fibrosis subsequently became in 1989 the first disease to become linked to a genetic mutation; others soon followed.

And now we have CRISPR—clustered regularly interspersed short palindromic repeats—and its star Cas9, a little enzyme that has a guide RNA at its core. There are bits of DNA between the palindromic repeats known as "spacers." Studies of bacteria resistant to viruses showed that the bacteria that had become resistant to viruses had actually incorporated bits of viral DNA into their "spacers." If you know an RNA sequence, you can search for the DNA spacer you want. CRISPR/Cas9's gene editing potential is already being researched for use in sickle cell anemia, muscular dystrophia, various kinds of cancer, stem cell use, and the possible inactivation of SARS-CoV-2. For their work in CRISPR/Cas9 genome editing, Emmanuelle Charpentier (French) and Jennifer Doudna (American) are awarded the 2020 Nobel Prize in Chemistry. Whether CRISPR/Cas9 proves to be a true scientific revolution or not, we have more women in the School of Athens.

When it comes to neuroscience, the history of medicine may have begun with the brain, but we still know less about it than perhaps any other organ. Like most of science, its history has been fraught with racism and sexism. One of the last shots of the "Apesh**t" video is of Marie-Guillemine Benoist's *Portrait of a Black Woman*, one of the pathetically few paintings of a brown body in one of the largest museums in the world.

CREATURES

Medical and popular interest in the neurological complications of covid burgeoned early on. We knew anosmia—loss of smell—as one of covid's early symptoms. It was questionable whether other neurological complications such as meningitis, encephalitis, strokes, and other acute presentations were due to normal complications of severe illness or otherwise specifically linked to covid. Popular news sources have been documenting case studies about long-term neurological complications and this has fueled anxiety in the imagination of the larger population. Months into this pandemic and we have patients who have survived covid coming into the emergency department, some transferred from outside hospitals, complaining of symptoms like acute leg numbness whose resultant history, examination, and imaging don't correlate to neurological manifestations.

There's a not-insubstantial overlap between psychiatry and neurology, and in fact practitioners of either become certified in a combined board: the American Board of Psychiatry and Neurology. When it came time to choose a specialization, I had narrowed it down to the two, but was uncertain until the end about which to pick. I had considered applying to a combined residency program in neuropsychiatry, but after researching it realized that you still ended up practicing as only one or the other, making the extended program superfluous.

Mental health stigma straddles the border between psychiatry and neurology. Epilepsy, characterized as a neurological disorder, has been equally mythicized and stigmatized over the history of mankind. The fact that anywhere from five to twenty percent of seizures are calculated to be psychogenic seizures does not help matters. From being considered the source of prophetic revelation, as when Dostoevsky in *The Idiot* refers to "Mahomet the epileptic," to a cause of debilitating persecution. Aristotle, Dostoevsky, van Gogh, Joan of Arc, and Dickens all allegedly suffered from either isolated seizure events or epilepsy. Although it is doubtful that this historical list is accurate, the very fact that such a list exists exemplifies the disease's perception. This trend of itemizing prominent figures who suffered from epilepsy or temporal lobe seizures has enveloped the disease with a sacred aura. Yet, by the same token, those with epilepsy have also been stigmatized throughout history. During the eugenics fad in the twentieth century, epileptics were persecuted along with various other targeted groups. Virginia's Sterilization Act of 1924 legalized the forced sterilization of people who suffered from "idiocy, imbecility, feeble-mindedness or epilepsy." President Woodrow Wilson, a fan of eugenics, supported this decision.

Yet this stigma is attached to the organ that we equate with our existence as individuals. The brain is the only organ with its own philosophical branch: philosophy of the mind. There's no analogous organ-projection equivalent. The closest would have been a soul emanating from the heart, but that's long since been debunked.

I grew up with both mind and brain. Both were a constant living and metaphysical presence in our house, from my mom's brain-shaped stress balls to my dad's boxes of neurological slides scattered in drawers around the house, to a library stacked with psychiatry and neurosurgical books end to end; from my mother's calming assurances that we were responsible for our own thoughts and emotions to my father's stem cell research into the recovery from traumatic brain injury. I thought I had had my fill of the brain growing up and it wasn't only until the latter half of med school that I came back around to it.

Between the two, I'm glad I ended up in neurology. I lack the temperament that makes for a good psychiatrist, and I had liked internal medicine too much to not get more training in it. But the mind is still my side chick and that includes the philosophy of the mind. "Physicalism" and "eliminative materialism" trapped and paralyzed my imagination in the latter half of med school. Between psychiatry's *Diagnostic and Statistical Manual of Mental Disorders* and neuroscience textbooks, I spent time on lecture series and podcasts on the philosophy of the mind.

The *Stanford Encyclopedia of Philosophy* describes eliminative materialism as follows:

> Eliminative materialism (or eliminativism) is the radical claim that our ordinary, common-sense understanding of the mind is deeply wrong and that some or all of the mental states posited by common-sense do not actually exist and have no role to play in a mature science of the mind. Descartes famously challenged much of what we take for granted, but he insisted that, for the most part, we can be confident about the content of our own minds. Eliminative materialists go further than Descartes on this point, since they challenge the existence of various mental states that Descartes took for granted

Seeking clarification, I turned to Noam Chomsky. I listened to his lectures on the cognitive revolution at the University of

Girona in Spain before reading *What Kind of Creatures Are We?* Both shed light on themes of language and consciousness. Language, according to Chomsky, is primarily a tool of consciousness at the conceptual-intentional interface, and not a result of communication at the sensorimotor interface, as some argue. Basically meaning that language evolved as an agent of consciousness and not as a tool of communication as people like to think. Chomsky makes a rather compelling and empirically detailed argument toward this thesis that can be found in the first chapter titled "What is Language?"

Chomsky is known as one of the founders of cognitive science and the father of modern linguistics, and in his lecture series he compares the twentieth-century cognitive revolution to that of the seventeenth-century. Both of these revolutions were sparked by similar factors, including the emergence of automata and computational theories of mind. The first cognitive revolution took the form of Cartesian dualism, considered to be on the continuum of Plato's Theory of Forms.

Chomsky's paper and a few others, including one by Allen Newell and Herbert A. Simon, contributed to the second cognitive revolution. According to him, the current cognitive revolution has taken a turn for the worse with the debate framed as "physicalism versus dualism." In an Open Yale course lecture, Dr. Shelly Kagan breaks the debate down through a semester's worth of knowledge, using the philosophical tool of "inference to the best explanation" to make his case. Bacteria, not spirits, for example, "cause" disease as a process of inductive reasoning. He claims a soul can only be explained if it survives this scrutiny. He debates back and forth, arguing that if there is something about the soul that physicalism cannot explain the argument goes to dualism. He calls human sensations *qualias*: tasting coffee, seeing the color red, experiencing life as a Pakistani-American, and the only differentiating factor between us and robots is this qualias. But it is only a matter of time before science explains qualias away, he argues; thus physicalism trumps dualism.

If we are just machines though, then there's no such thing as free will, belief, thought … leading to eliminative materialism. Chomsky has come to be seated outside this debate and see it descend into meaninglessness.

An example of the meaninglessness of this debate is the oft-cited idea that "consciousness is firing neurons" ("CIFN" to those in the know). By arguing against irrational dualism, the CIFN priests are actually framing and perpetuating it. Irrational dualism arises from choosing to study some subjects naturalistically and others non-naturalistically, meaning that the study of the mind, unlike the rest of nature, is not taken to have "limits and scopes." The closest example we have to the brain is the gut, which with its extensive neuronal activity is termed the "second brain" in medical literature. Chomsky uses this example to show that there is no "innateness hypothesis" about the gut. The gut, like the rest of nature, has limits and scopes, or, as Chomsky labels them, "problems and mysteries." Yet, it is taken for granted that the mind can understand the nature of reality, when all of nature shows us to have limits and scopes. The human mind can do certain things, because it cannot do others. Dogs may be able to see things differently, just as bats are posited to in Thomas Nagel's seminal "What Is it Like to Be a Bat?" Scientists such as Einstein have questioned how the human mind can be capable of understanding the nature of reality; how it was that from the Enlightenment onwards science was able to make such huge strides.

The problem is that the intelligible source of the universe is inside us. Chomsky says the onus of justification would fall on those claiming that something should not be studied naturalistically, in this case those who claim the human mind is capable of grasping reality—of jumping from subjective consciousness to an objective reality. The justification for this is often cited as evolution (by Charles Sanders Pierce and Stephen Hawking), but Chomsky cites leading evolutionary biologist Dick Lewontin in claiming that there is nothing in the evolution of humankind that would have naturally selected consciousness to understand reality, i.e. the "science-forming faculty."

We cannot firmly base the science-forming faculty, this intelligibility that has allowed us to understand the nature of reality. It could be wrong, it could be right, but it comes down to an unknown cause within the mysteries realm of the mind. In other words, within the problems realm is the mind's ability to experience, experiment, hypothesize, but within the mysteries realm is ultimate explanation. In his book Chomsky narrates:

> Historians of science have recognized that Newton's reluctant intellectual moves set forth a new view of science in which the goal is not to seek ultimate explanations but to find the best theoretical account we can of the phenomena of experience and experiment." As Richard Popkin claims, science proceeds by "doubting our abilities to find grounds for our knowledge, while accepting and increasing the knowledge itself," and recognizing that "the secrets of nature, of things themselves, are forever hidden from us.

We still have a long way to go with neuroscience. So to make grandiose claims about the nature of consciousness does not make sense. Mental states may be reduced to the organic function of the brain (both properties of this world and thus reducible in that sense), but we cannot claim to know more than that.

Chomsky relates that when Newton discovered gravity, his rival Leibniz claimed that it was occult, and Newton replied saying it was not occult, "their causes only are occult," leading to the famous statement, "hypotheses non fingo." It is frightening to face naked unintelligibility. Claiming ignorance and admitting to the mysteries of the world takes wisdom and courage, qualities that initially sparked the Scientific Revolution and that, unfortunately, parts of the contemporary scientific world have forgotten.

PANERA

People have lists of things they miss most about normal life during the lockdown, and mine includes coffee shops. I first started experimenting with automatic writing in a Panera Bread café, to commemorate my undergrad English classes. I would usually get coffee from the Panera on the Ann Arbor campus before going to class in Angel Hall, and had never wanted it to end. *Daniel Deronda* and *A Pair of Blue Eyes* and *Jane Eyre* formed my life's bread and butter in that age, enough so that I still wonder what it would have been like to have made it ageless. I'm mostly glad I didn't: better to un-passion a passion than to make a cliché out of it.

At the other end of automatic writing you delete all the fluff, including adjectives, adverbs, and prepositions. I learned my propensity for adverbs and prepositions through that exercise: I can't pretend as if I knew the lack before we knew what it was. I know that we left it long enough ago for it to no longer make a difference. Love yourself and what if the word doesn't exist in your vocabulary like Alain de Botton or Joan Didion; what if I just continue calling it butterfly or butter or whatever de Botton refers to it as in *The Course of Love*? I read it too long ago to clearly remember (but I'm proposedly in Panera and have already spoken of bread and butter). So, what do you do? Intellectual humans have told us the word must not exist, so it must not—except for the Sufi saints of an age ago who say it's the only thing to live by. I can still live by butter though, melted straight into our eyes and skin and pores. Here's the buttered toast and buttered croissants and buttered waffles and the pack of butter my baby cousin ate in whole after he'd been set on my grandmother's dining table at sehri.

Butter in the hospitals, where we learn to put it ahead of transference and countertransference and in the face of displacement and people crying on phones saying they can't afford funeral homes because they've been out of work in the covid era and the morgues are too full and still, for butter, we may go on, through the greater butter of humanity we can still strive toward (with another euphemism for God, for that has been taken away too, let's euphemize with coffee—) we may go on, for the greater butter of humanity we can still strive toward coffee and come back around to butter and to ultimate humanity (humanity still a permissible term in post-modernism).

We've had more deaths than Italy, and here we were joking about Italy's position in the EU. Silly us. How long will this last. I should walk in the sun if it weren't so cold. Her temperature had dropped perilously the night before she died, so cold the thermometer wouldn't pick up her temperature and we'd had to run to the OR to get a bear hugger.

I imagine us where we were before: has the image of my old age changed? What was old age when we were younger? If you think back of it from eighty what do you see? In order to see meaning, what must you see? A life lived gracefully and lightly and beautifully. A connection of webs with enough friends to have called home. What about when your friends stop being home and become distant enough for you to have to re-create them from a past recollection of home? How does that happen? Connection emanates from hope which emanates from reality. I'm forgetting how to be mortal which is stupid enough on its own. You have to butter this life before you want eternity.

I still haven't learned how to lose myself and it's starting to show. Embrace your own story and then delete it. Like, delete, dislike, delete, like, delete. But tumble down carefully. How low can you practically go? If you go too far in the process of negation you end up getting rid of all ideals, including the ones worth fighting for. What happens to systemic justices that arise from some form of idealism, with that process of negation? Leave some places for addition: fos + phenytoin, better for absorption that way, seizures better aborted. I worked and studied today and did more than other days and had a virtual iftar with my family upstairs from my isolated basement. Halfway through reading a poetry collection before going to bed the day had become unbearably pregnant.

One melatonin, two melatonin, three melatonin, the sheep bleating below the covers saved for sacrifice. I'm still in bed. Outside the Lahore hospital steps with Sara on a night where the stars could almost be seen, when I must have made some kind of silent oath that this wasn't real. I need to be rid of this story. A few mornings or months later or earlier I woke up at my grandmother's house to drink her famous honeyed frozen coffee, the summer air in the morning hall paltry and full, my company a dead butterfly on the large dining table where my cousin had eaten that whole butter in full over two decades previously—when early dawns with the fajr azan spirited us into existential flurries, splaying my cousins and I into my grandmother's room, Abdul Basit Abdus-Samad's crackling cassetted voice reaching all the way back. It's not real here, but it was less real there. Choose between the realer of your illusions. You became a part of my lingua franca—I told everyone. Tomorrow I'll learn how to love the morning and leave the night in peace.

MAY

"She begins by saying that we are having a wretched May, and, having thus got into touch with her unknown guest, proceeds to matters of greater interest."
—Emily Temple, "Essential Writing Advice from Virginia Woolf"

If you can't remember an event as it unfolds you can't remember it afterward. We hardly lifted the dust, for all it was worth, and only the less seemly changes have lingered. The economic crisis had been foreseen early on; its competition with health in full force throughout most of this and ongoing. I haven't seen some of the people I work with in months, and even when I have, they've been pairs of eyes floating over nose bridges.

It's been strange to realize how much of a face exists besides the eyes.

A restfully awake, closed-eyed human's EEG demonstrates a pattern known as alpha waves that are mostly omitted from the visual cortex in the occipital lobe. But with eyes closed, our faces, from outside in, are more than the visional plane of things. The Fusiform Face Area in the temporal lobe of the brain is one of the main areas attributed to facial recognition, and I wonder about the face's topical arrangement there. I've been surprised that the faces in my head don't match the mask-suffocated pink facial realities at the end of the day: some noses smaller and more angular than I'd imagined; some softer with more character. Like a focal deficit revealing functionality (in neurological franca lingua), I've realized the most defining feature of a face is the nose. Optic nerves may be direct extensions of the brain, but it's the nose that's the window to the soul. When I think back on Mr. Rochester, I am reminded of him by that holiest appendage: I recognized his decisive nose, more remarkable for character than beauty; his full nostrils, denoting, I thought, choler.

Moving nose onwards, the body has limited space for both ego and soul. Whether the body is in the soul or the soul is in the body, or there's really no "in" and "about" as it goes, I don't know or care to explore anymore. Being alive is a state of mind—in this version the guillotine drops earlier, the hare wins. Or let's not. When I wrote back to a friend on my socially isolated birthday that they lifted my soul, I wasn't talking about the Grim Reaper's way of lifting souls; I meant sinking souls. Souls engulfing flesh to their own design and centrifuging passionate egos out through pustular pores to settle at dermal fringes of the nose (hopefully vitamin C serum, dermatology's product du jour, helps that process as well).

In the past few months, my body has gone from living with a co-intern to living at a hotel to now having isolated myself in my family home's basement. But these external changes haven't (at least as of yet) made their way deeper into "thoughts and themes." That is, the idea of "change" itself hasn't made its way deeper into either medical culture or the culture at large. The break in the monotone existence of eighty-hour work weeks as a first-year neurology resident was somewhat of a reprieve; the rigidity loosened for a moment. Although the pandemic came with its own set of anxieties, I finally felt human after a long time—outside the constant fight for time that characterizes U.S. contemporaneity.

The focus on time as a primary enemy had been shifted to more pressing matters. It reminded me of the Paris I visited before starting residency. The streets at night, even on weekdays, even after midnight, were softly alive—something I hadn't seen even when I'd lived in New York. There, in Paris, was a culture outside the structure of weekdays and weekends; it wasn't stridently making up for lost time of weekdays, instead it was a man gazing out from a wrought-iron table outside a bar at 2 a.m. on a Thursday while drinking a nightcap—berried softly at a forefront of banlieues. Sleep and leisure were not scarce commodities bandaged over with scheduled "wellness" activities that nationwide most residents mock. Time was less rigidly demarcated and fought for there. The surrealism of the pandemic triggered forth a similar time-felt.

Things are reverting though. Clockwork clicking. The "Diabolum," as Milan Kundera named it, back in full force, the emergency department filling up with motor vehicle accidents once again. Retreating from the sickly self-hero worship, we've gone back to treating coding documentation instead of humans; words used to describe bodies justified over the bodies themselves. Call this a lack of caring or lack of enough momentum to have crossed over the barrier. More Americans have died during this (the echoed stats: a hundred thousand in three months) than the Korean or Vietnam wars, but let's not forget Europe had to decimate the world a few times over before the idea of human rights evolved. Maybe there aren't enough dead, maybe not in the way that matter, maybe postmodernity is too fossilized in its ways for anything anymore to really matter, or maybe time will prove me wrong as I'm watching protests against George Floyd's death flood the country.

My senior—the superior in residency who I, as a first year, am buddied up with—was going over a systemized approach to strokes and cerebral angiograms during our overnight call at the hospital. He broke off to take a nap, and I resigned myself to being unknown in memory—segments of the vertebral artery that winds up to supply the brain, V1, V2, V3, V4, in named blood and bone grooves. Microscopically moving closer, flipping memory upside down into eternity. V1-V2-V3-V4—where there's a group there's a way. Thirty-three vertebrae and you're the only one that matters—C7 jutting out as unforgivingly at the back of our necks as C2 is in hangings. If you think everyone else is hurting this much, 15L oxygen proned onto their bellies—or outside this realm altogether and still this much—then you must remember they're not and maintain that with dignity. I'll deal with the months of dreary brain for that imagined moment of reinstated large social gathering outside the Detroit Opera House in the parka-cold, a cigarette dang-a-ling in a distantly foreign dark hand.

Per Virginia Woolf, "Better was it to go unknown and leave behind you an arch, than to burn like a meteor and leave no dust." There is power in gentle digression, in talking a subject matter its way to silence instead of decorating said subject matter with blood and confetti. Authenticity speaks from a place of silence: an arch born from a bowed head. I'm curious to see the standing edifice of this pandemic once the blood and confetti settles. But for now, the world is on caution for either their first or second waves. Silent before and after who's to say anything even happened except for the dead, and they're no longer with us to provide an opinion. I'm hoping that they'll be spoken for, that some pus—or, alternatively, cold sweat foam—will be relieved via some volcanic orifice or policy (as long as we're still here on earth) or something. I'm as well as I'd ever wanted to be though, nose poked out to the window-veiled sky in this winter that's extended into late May in Detroit.

JUNE, AND ONWARD

THE DREAM OF A RIDICULOUS WOMAN

1

You would think cholera would be an old excuse for cause of death in the twenty-first century, but it isn't. Diarrhea is still a leading cause of death, especially in infants, in many so-called third-world countries. One of my maternal aunts, what could have been an aunt, died of it in infancy. In the first episode of the Bill Gates' documentary series *Inside Bill's Brain*, Gates reads a *New York Times* op-ed about how diarrhea still impacts the world today. This sets Gates on the course to find a solution.

Contaminated water due to improperly disposed human waste is the root of the problem, so he starts there. He holds a contest to find an economically viable solution to the problem of human waste and pit latrines and at the end of the first episode we have the solution in the Omni Processor, an array of treatments that both remove pathogens from human fecal waste and generate energy.

We solve problems only to be replaced by new ones, and you have to set aside cynicism to still go on solving problems when you know they're only going to be replaced by more and more complex ones. If you start doubting the myth of perpetual human progress the cynicism kicks in a square notch or two.

One of the many achievements of which modern medicine can boast is the decline in maternal mortality rates since the

1800's. However, despite this undeniable success, there's still a massive disparity in maternal and infant death depending on where in the world you're having a baby. How preterm you are in different parts of the globe dictates your probability of survival. Premature survival rates hauntingly dismal in the developing world. The probability of an existence is dependent on space. If existence is a matter of probability, then our personal stories are even more so just stories of probabilities.

2

Dear Ma, remember when I left the first time and it broke everything? I tried to come back but never really could. Remember when you taught me how to oil paint in the basement? I could only see the world in monochrome, and you added color.

Embryology was one of my favorite medical school courses—a three that for a suspended moment outside of time and space exist as one, when the spermatic egg and oocyte fuse to become enveloped in a membrane. They become one in a cataclysmic reaction doused in wavelengths of light seen and unseen by human eyes: red, orange, yellow, blue, violent, indigo, spineless bamboo, clear skin, gamma guts, tattered peaches, ultraviolet teeth. A Big Bang within the female's Fallopian tube. I had contemplated going into womens' health before I realized the practicality of it didn't quite match the textbook part that I admired.

I had expected a textbook C-section to be a legitimate alternative to natural childbirth, until I saw it firsthand. In a way it was almost worse: the ripping open of layers to get to the uterus to pull out a vernix caseosa-covered physiological entity. Surgery is well controlled, everything covered except for the markings of incision. Blood kept under control (unless you bust the wrong blood vessel and the patient starts hemorrhaging). But childbirth is its own world, gushing waterfalls of blood and bodily fluids. I have never felt faint before, but during my elective in Detroit I was helping close a vaginal laceration in a thirteen-year-old who had just vaginally delivered a baby boy, blood pooling at my feet, when my vasovagal response tremored, blood pooling. "Are you okay?" I'd nodded assent to the attending who'd asked but continued feeling lightheaded and asked to step out; one of the nurses got me a glass of ice chips as I sagged down a wall.

3

Lahore had been another cultural epoch. A young wife had had a difficult birth, was still screaming while the doctors tried to control the peripartum hemorrhage due to a placenta accrete (a placenta that's abnormally attached to the uterine wall). Through her bleeding core she was yelling, asking if it was a boy; her family outside was yelling asking if it was a boy. The infant was rushed to the neonatal ICU, and the pediatrician whispered, ambiguous genitalia. What would happen to this child? Would it be left on the steps of Lahore's khusra community? This family that was causing a ruckus up and down the hallways, demanding a boy as if this were some kind of mail order service. What would happen to the woman if she survived the hemorrhage? What would this family do to her?

Obstetrics/gynecology is practiced in a well-documented difficult environment. I was surprised to see how much was similar in both Lahore and Detroit. The doctors one step above surgeons as far as the cases are concerned, with all that gushing, called in at any time of the day or night. That environment is the result of too many strong-willed individuals coming to play together. I can imagine it being similar to a corporate boardroom of alpha males. If you've got enough willpower though, it's one of the most rewarding fields as well. The little crying neonate that comes at the end of it the closest thing to a miracle.

4

If history is the story of possibilities, one possibility stands out in clear relief. The TA of my undergraduate Roman history class was going over the First Council of Nicaea. Constantine convened the council in 325 A.D. in Nicaea, which is now in modern-day Turkey. The council's purpose was to unify Christian beliefs for the empire. The TA asked, mostly rhetorically, what would have happened if Arianism had won. I raised my hand: "It would have looked kind of like Islam." Arianism had held that Jesus was begotten and therefore lesser than God. Instead Homoousion's Holy Trinity overwhelmingly won the council.

This is just a single example, but it showed how interconnected everything is. If ends come with a breach, there's a possibility to lace them together into a mosaic—to go back to the point of branching. There's connection in division and that's a beautiful concept. It's sometimes hard and bloody and seemingly impossible, but if it weren't difficult you wouldn't need hope in the first place and then where would we be left with beauty? I'd been entranced by the Holy Trinity (its theological debates some of the most intricate out there) and about how much overlap there is between Judaism and Islam, not just the prophets, but in structures of law and mysticism as well.

There's so much overlap in everything that it can't be undone. History isn't only a list of possibilities, it's possibilities stacked on top of one another and melded at random spots (a tree spewed out). It needs to be taken with a grain of salt. You learn how anything we think of could be something else, but it still matters so damned much in the now. And it's okay that it matters in the now, it's just the intensity at which it matters that needs to be monitored through an occasional bird's eye view of things.

5

A part of adolescence is expanding beyond your ego and learning to live in statistics. As a physician it ends up being a game of death statistics with the Kaplan-Meier estimator. The statistics show that this is likely to help in this patient and that this isn't. Whether it does or doesn't is just reality's way of trouncing us every now and then. But we follow the statistics. Beyond that, things are out of our control. Which makes us more mechanized than anything else. If technology ends up replacing medicine, Yuval Harari in *21 Lessons for the 21st Century* made the point that it will be much easier to replace physicians than nurses. (Though with EMS systems reminiscent of Windows 95 I wonder how that feat will ever be accomplished.)

But for now, the superego is necessary. We need to feel purpose in what we do on a day-to-day basis. Physician, nurse, physical therapist, physician assistant, researcher. The very people driving any semblance of progress in this microcosm have strongly embedded inter-subjective realities. Whether it's a belief in the power of medicine, the power of science, the power of God, the power of nature, or the power of love, some mythical basis for meaning must be preserved otherwise the work would be impossible. The ends justify the means. If no myths result in a paralyzed human in the corner, an underground man, then is that really truth? That being the case, is it really even just a myth anymore?

6

There are two main attacks against meaning created by God: either science explains the world, no need for a God, or God is masked under ideology. The first is a less obvious problem to me; infinite regression not a sufficient reason to believe or disbelieve in God. When we think of God it's in a visceral, experiential way. It's the second that is more disturbing, but once you work on unveiling the unveiling process things become clearer—granted the acrobatic thought process that that requires. I liked how Paul Kalinthi formulates it in *When Breath Becomes Air*:

That's not to say that if you believe in meaning, you must also believe in God. It is to say, though, that if you believe that science provides no basis for God, then you are almost obligated to conclude that science provides no basis for meaning and, therefore, life itself doesn't have any. In other words, existential claims have no weight; all knowledge is scientific knowledge. Yet the paradox is that scientific methodology is the product of human hands and thus cannot reach some permanent truth.

Paul Kalanithi's story, of how a cancer diagnosis at the end of his neurosurgery residency forced him to re-evaluate what mattered in his life, of scraping together a distancing marriage and having a child, seemed to follow the general thread of wanting something ideologically and then reorienting to friends and family in the face of death.

7

It was impossible to run away from chaos growing up with a large Pakistani extended family, where the gaps between generations and socioeconomic classes are deemed secondary to togetherness. The chaos arises from different values, from atheism and religious doubt to religiosity and mysticism—one family member questioning others' belief in God, another taking us to Data Darbar's (a Sufi saint shrine in Lahore), another taking down family pictures for fear of blasphemy. There's a similar variety in political beliefs—family members who are government secretaries and ministers to followers of populist leaders to fringe believers in rebellion. Conversations turning into debates with pro revolutionaries pitted against anti revolutionaries, claiming the only thing the French Revolution succeeded in was bringing about Napoleon and plunging the world into WWI. Thereupon a disavowal of revolutions and a declaration that the American Revolution wasn't a revolution but a war of independence and a category unto its own altogether.

Then there was my grandmother's house, where a kitchen one day was turned into a living room the next, just when your edges started crusting from habit, boom, broken, keeping your spirit fluid. Our lives weren't bordered off by walls—houses and rooms were open for people to walk in and out and treated as shared space.

It's one of the many ironies of life, especially the older you get, that you can do more for others by maintaining both physical and emotional distance. Cultural dynamics have also shifted with technology. Expectations have become more consistent with nuclear families and extended family networks are starting to dwindle, the world inexorably moving towards a mostly global culture. Time only becoming an ever-greater reality.

Maybe this is the progression of stagnant thinkers: we work in reverse—from death probabilities outward. So, what happens when a stagnant human tries to become anything? We crawl man, we crawl.

POST-COLONIAL

The Big Bang banged 14 billion years ago; organisms popped up 4 billion years ago; animals like us, fellow *Homo* but not *sapiens*, danced onto the scene 2.5 million years ago. Our genus started using fire 300,000 years ago. *Homo sapiens* appeared on the scene around 150,000 years ago and 70,000 years ago *Homo sapiens* started spreading from East Africa into the Arabian peninsula (where something or other happened with Neanderthals) into Eurasia (where something or other happened with *Homo erectus*)— 70,000 years of human history and you're a part of it and I'm going into Yuval Harari's *Sapiens*.

Harari's Cognitive Revolution occurs 70,000 to 30,000 years ago and it's the appearance of language which allows humans to "bypass the genome." Whereas the behavior of other social animals is in large part restricted by their DNA, the same is not the case for our species because of human language and the myths that arise from it. That's not to say other animals don't communicate, of course they do, however, he makes the case that common chimpanzees live in societies with alpha males and the closely related species, bonobos, is ruled by female alliances—however, they wouldn't take political theory classes or lead feminist revolutions, behavior transmitted by communication and not genetic replication.

Using that logic, he slightly undermines the evolutionary psychology that limits our psyche to a hunter-gatherer brain. Even pre-agricultural hunting-gathering tribes were so distinct from one another (at least from the limited knowledge we have of them) that to lump them into a singular "natural way of life" misses the point entirely: language freed us from behavior that arises solely from DNA that dictates other social species. He writes:

> The Cognitive Revolution is accordingly the point when history declared its independence from biology. Until the Cognitive Revolution, the doings of all human species belonged to the realm of biology, or, if you so prefer, prehistory.... From the Cognitive Revolution onwards, historical narratives replace biological theories as our primary means of explaining the development of Homo sapiens. To understand the rise of Christianity or the French Revolution, it is not enough to comprehend the interaction of genes, hormones and organisms. It is necessary to take into account the interaction of ideas, images and fantasies as well.

Just as there's no "natural" hunter-gathering brain; there isn't an untarnished pre-imperial culture (humans have lived in some empire or another since at least 200 B.C.). Money, the root of evil, is also a universal bridge between different religions and languages, and liberalism, communism, capitalism, and humanism are on the same threshold of ideologies as religion. Harari's deconstruction isn't new, but it's beautiful how he transitions from one theme to the next so seamlessly and objectively. He even talks about the differences between objectivity (an apple will fall toward earth whether or not someone chooses to believe in gravity), subjectivity (the imagination of a single human, like an imaginary friend), and inter-subjectivity (shared myths, from money to economics to religions). He's cognizant that a great deal of his subject material falls into the inter-subjective domain and avoids value judgements.

Harari spends the most time on the Scientific Revolution, so intertwined with modern empire building that it's difficult to separate the two.

In a chapter titled "There is No Justice in History," he argues that a lot (if not most or all) of empire building can be considered evil, especially the lingering racism that has been replaced by "culturism." Similar idea, different -ism. It's politically incorrect to dismiss an contradictory behavior as racial, but more than acceptable to say "it must be a cultural thing."

A lot of the medical revolutions or humanist ideals that came out of colonialism were good, but to label one evil and one good when their existence depended on each other leads to inconsistency, especially when those colonized peoples adopted the very ideologies that their colonizers had used to prosper later on. This makes the decolonization project contentious:

> There are schools of thought and political movements that seek to purge human cultures of imperialism, leaving behind what they claim is a pure, authentic civilization, untainted by sin. These ideologies are at best naïve; at worse they serve as disingenuous window-dressing for crude nationalism and bigotry.

But, to play devil's advocate here, inconsistency is also the basis of what makes us human, as Harari himself writes in the first part of his book:

> [C]ontradictions are an inseparable part of every human culture. In fact, they are the engines of cultural development, responsible for the creativity

> and dynamism of our species...Had people been unable to hold contradictory beliefs and values, it would probably have been impossible to establish and maintain any human culture.

Even when we are calm and in love and urban, out shopping at the local Kroger, a part of it is all still pretense. Especially because of those fluorescent lights bouncing off waxed surfaces that have never seen dust.

Once an inconsistency is realized it's prudent to take a step back and reframe what it is that you're believing or doing. European imperialism sent all kinds of specialists and scientists to the conquered lands, the most famous example being of Napoleon's expedition to Egypt. The better they knew the lands the more successful the conquest. It was the unprecedented greed of the modern imperial project that fueled the Scientific Revolution. Harari calls it the military-industrial-science complex. "Science research can flourish only in alliance with some religion or ideology. The ideology justifies the costs of the research. In exchange, the ideology influences the scientific agenda and determines what to do with the discoveries." This is the complex that exists even into today. Approaching this complex invariable requires the ability to juggle contradictions and paradoxes.

Harari only brings up value judgments at the end of his work. Was it all worth it? All these revolutions and ups and downs, with the Scientific Revolution on the brink of changing the very mechanism of evolution—genetic engineering possibly tipping us into a new cosmic era as he calls it. What would the measuring stick of worth look like? If we define it as happiness, happiness is either framed by the biological serotonin theory, which quickly takes you down the *Brave New*

World path, or happiness is a form of delusionary meaning. In the former there's a want for pleasant sensations and in the latter a want for meaning. But, Harari writes, both are simply wants, and based on subjective feelings unique to liberalism, the pursuit of happiness, whereas throughout history most religions and ideologies have used objective markers for ideals for goodness and beauty. He states that feelings were viewed suspiciously, and writes that the "Know thyself!" inscription at the entrance of Apollo at Delphi was a reminder of the average person's ignorance to their true feelings and in fact ignorance of "true" happiness. Instead, happiness or meaning ensued from feelings that were subject to morality. The modern day's elevation of sentiment has made that definition obsolete.

The alternative to the pursuit of and addiction to feelings he offers is found in Buddhism. Not the New Age version that has been adopted into the dominant liberalistic rhetoric, but the "original one" that not only undercuts external achievements but also defines a happiness outside subjective feelings and wants. Harari writes that the problem isn't suffering or meaninglessness; the real problem is the pursuit of temporary feelings.

The solution to meaninglessness isn't to search for meaning. The solution is to realize the inherent flaw in the quest itself—in elevating the ideal over the real. Meaning is the hope that naturally emanates from the curiosity and acceptance of reality, whether that includes suffering or no suffering or life or death. Paradoxically, meaning is found outside of meaning.

CRAWL ME A SLICE

It's finally warm outside. We made it through our intern year, the interns who lived through the pandemic, which I'm sure will make for a good story towards the ends of our lives when we're wrinkly and hooked up to dialysis machines, with baseball caps alluding to the earlier part of this century. A new intern (though I hope there's something better than interns by then) will be surprised to find a human who's lived through something only read about, like the way we look at the rare human who made it through the Spanish flu.

July marks my transition into my second year of training. The epicenter has moved from the Midwest and Northeast to Florida, Arizona, and California. The Plaquenil debate is still continuing, with most research indicating an end to its use in covid patients. CNN releases Henry Ford's anomalous promising findings associated with Plaquenil, pretty much to the consternation of everyone who hates Trump. America begins its withdrawal from WHO, and The *New English Journal of Medicine* publishes a report that antibodies against SARS-CoV-2 drop dramatically after mild infections in the first three months.

The spike protein, already having been researched in previous coronaviruses like MERS, is the antigen of interest in the vaccine race. The two mRNA vaccine contenders are Moderna/NIH and Pfizer/BioNTech. The viral vector vaccines are Johnson & Johnson and the University of Oxford/AstraZeneca. On July 14 Reuters reports the first-phase trial for Moderna, the first to enter large-scale human trials, as being both safe and effective. The acronym we had learned for the phases of clinical trials in med school is SWIM: 1) Is it safe? 2) Does it work? 3) Is it an improvement from the standard of care? And 4) Market. The U.S. signs a contract with Pfizer for vaccine delivery by January.

I'm here enough at this moment to make it into a scene: the decking tiles that I laid out on the balcony of my newly rented resident apartment ripped of wants to flood flood flood a meaning that can only be breathed into and out and then back into and out and never long enough to stay. Unlike the hotel room, my new apartment has a wide, sweeping glass door that opens onto a view of Detroit's skyline. Keep your heart close and your daughters closer. Kanye had a florid manic episode the same week Taylor Swift came out with a beautiful album, and I'm tiptoeing around the debate of loss of accountability with mental pathology, but I can't help thinking of this as poetic justice from TSwift's end—if it isn't right it isn't the end.

My second year of training is spent mostly on the stroke unit and taking consults. Things in the hospital have normalized except for all the masks and virtual didactics. While life for health workers has retained some level of normalcy, the majority of humanity's existence has been upended—my family and friends have had to work and study from home for months by this point. Why does this feel so abnormal and what was normal to begin with? It's strange that being able to walk out of the house means so much. But then again, all paths lead to movement. It's 11:11 with Khalid and Summer Walker's "Eleven" on repeat. I can't reproduce the lyrics here, but they're singing about late night cruising and watching each other's movements.

It's August and I have my first two weeks of night float, when we work through night shifts from 7 p.m. to 7 a.m. A stroke code is activated one night on a man visiting from Arizona. I examine his sweating, feverish body before I get a chance to put gloves on, and his covid test results positive hours later. The exhaustion starts kicking in by the third night, the lack of natural light contributing to a warped sense of time and existence. I haven't had time to buy a headframe, and the bright day outside makes it difficult to sleep on the mattress on the floor. The Tuft and Needle foam mattress had been delivered to my apartment mailroom wrapped in plastic like a burrito.

An opinion piece by Drs. Sarah Matathia of Massachusetts General Hospital and Monique Tello of Harvard Medicine in *Scientific American* (August 27) talks about the Flexner report. The report resulted in a rift between public health and medicine that 110 years later has made this pandemic worse. They write that public health measures, like wearing masks, testing, and isolation, are distanced from practical medicine and have been undermined by national politics. The first case of a covid reinfection is reported. Moderna announces its supply agreement with the U.S.

It's September and Ruth Bader Ginsburg dies and Trump replaces her before the rest of us can blink. She's the second woman in history to serve in the U.S. Supreme Court, after Sandra Day O'Connor. The Hulu documentary shaping this powerful five-foot-one woman's story highlights a few points: her living by her mother's message to "be a lady and not be distracted by anger," her life-long relationship with her supportive husband, and her famous friendship with the conservative justice Antonin Scalia. A September 21 *Literary Hub* article reminds the world of Vladimir Nabokov's influence on her writing and reading when she was a student at Cornell.

Pfizer expands the number of people participating in the phase three trials of their vaccine. According to *The New York Times* (September 28), the number of worldwide coronavirus deaths crosses the one million mark.

I'm on night float again for the first presidential debate on September 29. My senior and I laugh through what has to be one of the most ridiculous American presidential debates in history. One of the patients on the floor is admitted for a transient ischemic attack (colloquially known as a minor stroke) after his stress levels shot through the roof during the debate. I return in the morning to sleep on my recently bed-framed bed, grateful for the elevation from the floor.

It's October and while standing in line at the hospital Subway I check my Twitter feed to see that the Irish Supreme Court rules Subway's bread isn't actually bread. My night float senior shows me the tweet that Trump has tested positive for covid. He is hospitalized and survives. My family and I cast our mail-in voting ballots from home. I sit at the kitchen table darkening the rectangular boxes next to the candidates' names.

Time's "100 Most Influential People" October issue includes Dr. Zhang Yongzhen, who publicly shared the genomic sequence for SARS-CoV-2, Megan Thee Stallion, who publicly shared "WAP," and of course, Dr. Anthony

Fauci. AstraZeneca and Johnson & Johnson restart their halted vaccine clinical trials. Thirty-four editors from *The New England Journal of Medicine* publish an October 8 editorial: "Dying in a Leadership Vacuum" that starts trending within hours. It criticizes the Trump administration for having "taken a crisis and turned it into a tragedy."

It's November and election day and it's election week and the results are dragging on and it's the end and Joe Biden and Kamala Harris are in. Denmark culls seventeen million mink because of coronavirus spread. Pfizer releases its vaccine trial data claiming 90% efficacy. Moderna and AstraZeneca announce similar efficacies with their vaccines. The FDA approves bamlanivumab for high- risk covid patients.

The covid cases are starting to surge again. A new emergency order is put in place restricting social gatherings. ICUs are preparing again with the plastic partitions going back up. Steroids work well before patients are vented, but once patients are vented we have limited treatment options that actually work. In addition to masks, we have to wear eye protective goggles and I perpetually feel like Amelia Earhart. I cancel my plan to attend my best friend's wedding, something I never would have imagined. It's November and we've had our first snow.

I run into the new interns, faces I've never seen behind surgical masks, who still unadulteratedly empathize with their patients. As a second year, I'm jaded and perfunctory. A new intern sits at her patient's bedside, the patient's hand in hers, as she tries to calm him down from an obvious panic attack. I wait to finish my neurological examination as a part of the stroke code that was activated and get back to the rest of my rounding. I am surprised to remember that that was me only a year ago, when I'd followed a patient's re-admission to the hospice floor and placed my hand over his death-waxing heart. Dig in the heels, clap back. I have to purposefully slow myself down instead of rushing from one consult to the next. Time can come and go, moments stay though.

It's mid-December and the vaccine has been approved and shipped and our hospital has started vaccinations. "Vaca" comes from cow, and it's from the initial immunization of cowpox developed by Edward Jenner (1749–1823) to make people immune to the more fatal smallpox. Smallpox was also the first disease to be successfully eradicated in the wild by vaccination.

Events and months and now years are rolling out like coins. In her essay "Some Notes on Attunement," Zadie Smith describes Kierkegaard's attempt to understand Abraham's story (the one where God tells him to sacrifice his son, only to have the son replaced by a ram at the last minute): "Faith involves an acceptance of absurdity. To get us that point, Kierkegaard hopes to 'attune' us, systematically discarding all the usual defense we put up in the face of the absurd." The years in my life have started adding up as a damning tally and I feel caught between the absurdity of reality and wanting to feel protected, closer.

Some days I'm everlasting enough to make it all last. Most days I'm not enough to make it through that last blast, but most days most don't believe in life or death or death or life. They just make it through to the end completely endless—

that's my only goal really: to be endless to the end. Most days we fight against what we know as death with numbers and words. It'll take a minute to figure out what it all does, all that talk of no will no free no something; until then we'll just have to make space between words and thought, wedged open with a nudged gentleness moving through the night, trying to fill in the space ever after.

It's December 22 and my dad and I got inoculated with the first dose of the Pfizer vaccine. It's December and the UK has a new strain of covid. If it feels like the end it isn't the end.

WORKS CITED

Adamson, Peter. *Health: A History*. New York: Oxford University Press, 2018

"ATLAS Club Presents The History of Medicine Part 1: Ancient Times to the Renaissance." (November 6, 2020). Retrieved January 3, 2021, from https://www.youtube.com/watch?v=Z3y3Y61t6E8

De Botton, Alain. *The Course of Love: A Novel.* New York: Simon & Schuster, 2016.

Chomsky, Noam. *What Kind of Creatures Are We?* New York: Columbia University Press, 2015.

Frankl, Viktor E. *Man's Search for Meaning*. New York: Beacon Press, 2014.

Iqbal, Muhammad. *The Reconstruction of Religious Thought in Islam*. Stanford: Stanford University Press, 2012.

Harari, Yuval Noah. *Sapiens: A Brief History of Humankind.* New York: Harper, 2015.

Harari, Yuval Noah. *21 Lessons for the 21st Century.* New York: Random House, 2018.

Moore, Graham. *The Last Days of Night: A Novel.* New York: Random House, 2017.

Nagel, Thomas "What Is it Like to Be a Bat?" *The Philosophical Review*, Vol. 83, No. 4 (Oct., 1974): 435-450.

Sacks, Oliver. *The Man Who Mistook his Wife for a Hat, And Other Clinical Tales*. New York: Summit Books, 1985.

Shem, Samuel. *The House of God*. New York: Berkley Books, 2010.

Smith, Zadie. "Some Notes on Attunement: A Voyage Around Joni Mitchell." *The New Yorker*, Dec. 17, 2012.

ACKNOWLEDGEMENTS

Any human-made institution, including medicine, will always have flaws, but sometimes I wonder who thought it was a good idea to throw a bunch of humans who have basically spent three decades of their lives with their noses and minds in textbooks into a setting where they suddenly have to deal with humans on an intimate basis. You have to develop a new skill set virtually overnight. With time, though, you look back at this with wondrous humor, although I do wish that things were a little more distant from "The House of God" than they are.

Thank you to the seniors and attendings who gently helped me develop that skill set in circumstances that were often far from gentle. And of course, my amazing neurology cohort which I couldn't have survived without: Drs. Enrique Martinez, Natalie Stec, Rami Al-Hader, Mo Fraz, Stephanie Phillips.

My parents, Asim Mahmood and Noorulain Farooqi, and sisters, Nijah and Aamal Mahmood, who have been my biggest supporters and suffered the many ups and downs with me on my academic journey and without whom I wouldn't even be a human so there's that.

Khadija Bokhari and Houriya Mukhtar—it's a rare blessing having best friends who have grown with you intimately since childhood and with whom you share a reservoir of memory. Dr. Sadia Maqsood and Dr. al-Khayat showed me a kindness when needed. Martha Bayne, thank you for being the editor I so needed—the faith and work you've put into this has been beautiful.

Justine Castle Photography

ABOUT THE AUTHOR

Selina Mahmood was born in Detroit and serves as a second-year neurology resident there. She has also lived in Lahore, New York City, and Ann Arbor. She graduated with a major in history from the University of Michigan in a previous life before pursuing medicine. Her work has appeared in *The Manhattanville Review*, *The Shallow Ends*, *Squawk Back*, *Blood and Thunder—Musings on the Art of Medicine*, *The Conglomerate*, and elsewhere. She has also blogged book reviews on HuffPost and worked as a reader for *Boulevard*, *Bellevue Literary Review*, and *Frontier Poetry*. When she isn't busy diving into the brain, she's trying to swallow her way out of it.